SAVING THE WORLD

SAVING THE WORLD

How Forests Inspired Global Efforts to Stop Climate Change from 1770 to the Present

Brett M. Bennett
and Gregory A. Barton

REAKTION BOOKS

To the memory of Frederick J. Kruger,
a great ecological thinker,
mentor and friend

Published by
REAKTION BOOKS LTD
Unit 32, Waterside
44–48 Wharf Road
London N1 7UX, UK
www.reaktionbooks.co.uk

First published 2024

Printed and bound in Great Britain by TJ Books Ltd, Padstow, Cornwall

A catalogue record for this book is available from the British Library

ISBN 978 1 78914 874 9

CONTENTS

INTRODUCTION
THE FORGOTTEN HISTORY OF CLIMATIC BOTANY

Global efforts to control the Earth's climate began on a small scale. Mauritius, an island smaller than Luxembourg and located in the Indian Ocean, is known today for its tropical beaches and warm ocean. Yet Mauritius is also sadly known best around the world for something it no longer possesses. The dodo bird (*Raphus cucullatus*), which went extinct in the 1690s, only lived on that one island. Over time, the dodo adapted to the lack of predators in the forests on the island, and eventually the species became flightless. The dodo went extinct when colonial sailors and later settlers hunted the bird and introduced pigs, cats and dogs, which predated on dodo birds, chicks and eggs. The island's ecology changed dramatically as its land came under the control of four successive colonial governments, starting with the Portuguese at the Treaty of Tordesillas in 1494 to the British handover in 1810 before eventual independence in 1968. The island's strategic location between East Africa and India, combined with its arable soils and pleasant maritime subtropical climate, meant that its indigenous species and ecosystems, from the dodo bird to native forests, experienced dramatic, and often alarming, losses.

By the 1770s, the loss of native forests had so concerned one French official on the island that he instigated arguably the first attempt by the state to stop human-induced climate change. Pierre Poivre (1719–1786), the French colonial government's intendant of Île de France and Île de Bourbon (known today as Mauritius and Réunion Island) from 1767 to 1772, proposed a bold and sweeping plan to fight changes to the climate caused by human activity. Poivre argued that the destruction of native forests since colonization had caused a dramatic decline in rain on the small island.[1] When people cut down forests, he argued, rain-laden clouds passed over the islands without so much as giving up a drop of rain. Poivre created policies to stop deforestation and to replant trees to produce a climate of 'perpetual spring'.[2]

Poivre, like many naturalists in the late 1700s and 1800s, believed that forests played a central role in local and regional climatic regulation, a concept this book describes as 'climatic botany'. Much as people save native forests and plant trees today to lower carbon emissions, so too did people in the 1700s and 1800s conserve forests and plant trees to manage temperature and precipitation. Poivre and other early modern experts in climate did not worry about greenhouse gases as people do today because the field of chemistry was then still in its infancy. Carbon dioxide had only been isolated and described by Joseph Black (1728–1799) as 'fixed air' in 1754. It was not until 1896 that the Swedish scientist Svante Arrhenius (1859–1927) famously argued that increasing emissions of carbon dioxide would raise the Earth's temperature. The alarm that the Earth is warming rapidly because of increased greenhouse gas emissions only became the primary concern of climate scientists in the late 1970s.

Poivre's life experiences put him in an excellent position to understand mid-eighteenth-century theories of forests and climate change. He knew a great deal about tropical plants because of his extensive travels throughout Asia, first as a missionary, later on various plant-hunting expeditions and finally as the intendant for Île de France and Île de Bourbon. Poivre had an especially colourful life: he served time in a Chinese prison in 1745 for working as a secret Christian missionary in Canton, lost part of his arm in a naval battle and, most famously, smuggled the spices cinnamon, clove and pepper out of the Dutch East Indies to Mauritius at the risk of death in 1771. Many of his ideas on forest protection stemmed from his beliefs about the importance of land as the underpinning of the economy, a belief he held as a physiocrat, that is, an economist who followed the philosophy of physiocracy, a French school of economic theory.

His ideas and actions, and the consequences they set in motion, foreshadowed two and a half centuries of further debate about whether human action can change the prevailing climate and, if we can change it, then how the state and civil society should respond. Colonial plantation owners wanted to grow spice trees and they saw any threat to the expansion of farms as a threat to their profit and viability. Poivre, to the contrary, believed that deforestation threatened the entire agricultural economy. He noted that there was abundant rainfall in forested areas of the island 'whilst the cleared lands are scarce watered by a single drop'.[3] There could be no agricultural economy without rain.

Like today, historical debates about climate change raised bigger questions about how humans should live, the necessity for government regulations and the limits of personal freedom. That same year, the French naturalist Abbé Richard (b. 1720) wrote in his book *Histoire naturelle de l'air et des météores* (Natural history of air and weather) that the study of weather and climate 'was not merely a speculative study' but that it also proved 'useful in the broad scheme of governing men'.[4] Scientists and officials as early as the 1760s and '70s began to propose government policies to reverse and prevent the consequences of human-induced climate change. The argument that only scientists and the state, rather than individuals and the market, could safely be counted on to conserve the climate is at least 250 years old.[5]

Unfortunately, Poivre's vision of planting a verdant Eden on Île de France failed to take root. The reforms possibly slowed the rate of deforestation while he was in office, but the island's government – like almost all early modern colonial governments – lacked the ability to closely regulate privately owned or controlled land owing to a lack of authority and staff to enforce laws as well as poor transport and communication. In the centuries since Poivre's reforms, the island steadily lost its native forest. This trend has only stopped within the last fifty years, once there were no longer any unprotected native forests to cut down. Today, native forests cover only 2 per cent of the land in Mauritius, and they are located primarily in government-run nature reserves.

In many respects, Poivre's efforts to stop human-induced climate change happened a century too early. Many naturalists in the late eighteenth century believed that forests regulated climate, especially in tropical regions, but this idea did not become of actionable concern among the public or political elite in Europe and its colonies until the mid-1800s. Deforestation was the most discussed and acted-on climate issue relating to human activity in the second half of the nineteenth century. At the peak of its popularity, adherents warned that deforestation would cause a devastating chain of events that could turn once fertile forests with ample precipitation into arid deserts with poor soils. Climatic botany helped launch the world's first environmental movement in the mid-1800s, well over a century before scientists started publicizing their warning about global warming.[6]

Today, few people know about this fascinating history even though it has significant bearing on how we respond to the various climate and environmental crises we face. The movement to regulate climate using forests – a movement that has existed for more than two centuries – has unfortunately not been studied by scientists as part of societal responses to mitigating contemporary climate change. The environmental historian Meredith McKittrick argues that older ideas of climate change 'are regarded as historical dead ends of little consequence, aside perhaps from helping us understand the context of the times that produced the fields of meteorology and climatology'.[7] The progress-focused nature of scientific enquiry, combined with the popular belief that the contemporary climate crisis is without precedent, all serve to downgrade history as a tool for understanding the present and future.

Rapid and exciting developments in scientific thinking on how forests influence precipitation and climate make this history even more relevant. Within the past decade, a growing number of water and climate researchers have argued that the influence of forests on the climate – from generating rain to cooling temperatures – is fundamental for climatic stability at local and regional scales. In 2017 a team of scholars led by David Ellison wrote that 'forests also provide a broad range of less recognized benefits that are equally, if not more, important [than carbon].'[8] A growing number of self-described supply-side researchers believe that trees are essential to the recycling of moisture necessary for precipitation and climatic stability.

Research suggests that forests play important roles as climatic regulators and as generators of rain. Evidence indicates that the Amazon and the Congo Basin generate roughly half of their precipitation through a process of moisture recycling, which involves water moving from the forest back into the atmosphere, where the moisture falls once again as precipitation before the cycle repeats itself.[9] Deforestation in the wet tropics not only leads to increased carbon and biodiversity loss, but changes the precipitation regime, temperature, humidity and the reflectivity of solar radiation. Wet rainforests are not the only places where recycling determines precipitation. Upwards of 80 per cent of the precipitation in Central Asia, including in western China, comes from water vapour from the Atlantic Ocean that has been continuously recycled from vegetation as it moves from west to east.

The supply-side view presents a challenge to its counterpart, the demand-side viewpoint, which emphasizes that environmental management policy should focus on conserving the water that already exists on the Earth's surface, such as in the form of lakes and streams, rather than trying to increase precipitation. These ideas gained prominence after the Second World War when studies showed that an increase in tree cover in a catchment led to declines – sometimes significant ones – of water within the waterways of the catchment area. As a result, demand-side policies restrict the planting or growth of trees in catchment areas in order to increase the availability of water because research shows that trees use more water than smaller vegetation types, such as grass or heath.[10] Ellison, an important scholar on supply-side hydrology, has spearheaded a scientific 'call to action that targets a reveal of paradigms . . . that treats the hydrologic and climate-cooling effects of trees and forests as the first order of priority'.[11] The supply- versus demand-side debate has profound implications for how we manage water and forests in the upcoming century. Current demand-side policies limit tree planting in many catchments for fear of using limited water supplies, but they do not consider how trees recycle rain to downwind regions.

Scientists who study forests and climate in the present day know very little about the history of climatic botany. Policy documents and major scientific reviews on the revival of the idea that forests induce precipitation do not mention historical precedents. A 2012 report for the British Meteorological Office writes incorrectly: 'The possible influence of forests on regional rainfall was first suggested in the 1970s.'[12] Without doubt, researchers in the 1970s made findings that provided strong evidence of a link between forests and precipitation, but they hardly proposed this idea for the first time. Poivre, among many others, had touched on this very issue two centuries before. Contemporary supply-side scientific theories, though much more robust and detailed owing to the technology available today, offer the same basic physical explanations as older theories of climatic botany for how moisture is recycled back into the atmosphere. Both theorize that vegetation plays a significant role in the recycling of precipitation across a range of climates and ecologies, from wet rainforests to arid deserts.

Scientists and the public are unaware of the fascinating history of climatic botany, partly because there is no book that takes this history from its early modern origins – with the expansion of European empires and the Scientific Revolution – all the way up to supply-side research in the present day. Academic historians have published many important books and articles on the history of forests and climate, but the majority of the scholarship is so specialized and fragmented by time periods and region that it would require becoming a professional historian to make sense of it all.[13] Many histories of climate change mention ideas of climatic botany, but none explore the subject in-depth or take the story into the late twentieth century.[14] Each chapter of *Saving the World* focuses on important events and periods from the mid-1700s onward

– including the revival of climatic botany since the 1970s, a topic that has not been explored by historians to date – to piece together this overlooked story of forests and climate.

Many supply-side policies enacted in the twenty-first century, such as planting trees and regenerating forests, resemble similar ideas tried in the nineteenth and twentieth centuries. Climatic botany has a rich history that must be studied with great care because forestry policies in the nineteenth and early twentieth centuries fundamentally restructured nature and peoples' livelihoods in ways that reverberate to this day.[15] In some cases, the creation of protected forests stopped the worst abuses of rampant deforestation from continuing. Tourists can today visit the world's largest trees, the Sierra redwoods (*Sequoiadendron giganteum*) in California and the mountain ash (*Eucalyptus regnans*) in Victoria, Australia, because forest protection laws stopped the forests from being stripped bare by rapacious commercial loggers. Yet these forestry laws also had serious downsides: Indigenous people and vulnerable groups throughout the world (especially in former European empires) have been dispossessed from their homelands and stripped of their livelihoods.

The management of forests by scientists has often tended to simplify ecosystems in favour of a handful of commercially valuable tree species, such as the Norway spruce (*Picea abies*), which was planted widely in Central Europe from the 1700s to the 1900s.[16] In many places, the planting of monoculture plantations destroyed rare and biodiverse grasslands, heathland and swamps. The history of climatic botany, specifically, and the history of forestry more generally, has practical relevance to scientists and policy makers working in this space today.

Climatic Botany from Pre-History to the Present

The idea that plants influence precipitation more specifically, and climate more generally, can be traced as far back as the ancient world. If we count traditions of rainmaking as part of this story, the origins stretch back into the misty dawn of the Palaeolithic era before humans developed agriculture and domestication. Proponents of rainmaking and climatic botany share the view that human activity, in one case prayer or magic and in the other tree planting, can influence precipitation, but rainmakers and climatic botanists differed on the exact mechanism that caused it to occur. Most recorded rainmaking practices were attempts by humans to harness supernatural powers to change specific weather events, such as bringing forth rain when it was needed, but not the climate itself over longer durations of years or decades. Climatic botany, to the contrary, proposes that the biophysical process of planting or cutting trees can change the climate over longer periods of time by either increasing precipitation, in the case of planting, or decreasing it, in the case of cutting.

Historians often point to Theophrastus (371–287 BCE), an ancient Greek naturalist and philosopher, as the first person to publish a naturalistic argument that vegetation could influence rainfall.[17] His ideas, which formed an important part of ancient Greek thinking about the natural world, did little to inspire action to protect forests at the time. There were no environmental movements in the ancient period, although prominent Roman writers did worry about arresting declining soil fertility caused by farming practices.

Like the Romans, many past societies have had some form of environmental reflexivity that allowed them to see environmental degradation, such as polluted rivers, as problems.[18] Yet we must remember that past societies employed different concepts, many of them unrelated to our modern scientific explanations, to explain the causes and meaning of changes in nature. The decline of soil fertility in ancient Rome could be explained as a consequence of bad farming practices, an idea that makes sense today, but it was just as often seen at the time as the consequence of a natural decline in the earth's fertility, a view that fitted with ancient Greek ideas of nature gradually declining in physical vigour – like humans do as they age.[19] The Christian belief that humans were banished from the Garden of Eden supported the view that the ancient world was better.[20] Religious discussions of the Flood played a major role in shaping European conceptions of climatic change in the early modern period at the same time as scientific conceptions of climatic botany were emerging.[21]

Ancient and later medieval European scholars worried much less about climate change than do their modern counterparts because, prior to the Scientific Revolution, scholars did not think that human action alone could dramatically change the climate or cause extinctions.[22] Claudius Ptolemy (100–170 CE), the most influential geographer of the ancient period, did not view climate as being something that changed because of physical causes, such as deforestation. His *Geography* argued that climate was determined primarily by the geographic location of a place rather than by the physical dynamics of its surrounding environment. Ptolemy also suggested that climate determined human characteristics, such as skin colour, culture and morality. Ptolemaic ideas of geography informed over a millennium of scholarly thought, and they remained popular among scholarly communities in Europe until the dawn of the early modern era.

With the onset of the Scientific Revolution in the early 1600s, naturalists began to see climate as an overall outcome of measurable natural forces – such as topography, temperature and humidity – in addition to a place's geographic location on the Earth's surface.[23]

Geographic location, though important, did not exclusively determine the climate. By the 1600s, naturalists understood that places at the same latitude, such as London in England and the region of Labrador in Canada, had widely divergent temperature ranges; this suggested that factors other than location alone determined climate. Early modern scholars of the ancient world noted significant changes to various regions' climatic conditions over time by comparing climates described in ancient texts to those of the same areas in the early modern era. For instance, descriptions of Lebanon in the Bible refer to a well-forested country and Roman depictions of Algeria portrayed it as a fertile breadbasket of the Mediterranean.[24]

Geography continued to shape how modern Europeans understood climate even after new biophysical explanations became the dominant views among scientists.[25] As a result, many Europeans continued to maintain into the twentieth century that hot, humid or extremely dry climates produced negative human characteristics, such as laziness or greed, while cooler climates produced more intellectual vigour. In 1909 the Australian Commonwealth government decided to create a national capital, Canberra, in one of the country's coldest places, partly because they believed the cooler climate was more conducive to Europeans and the intellectual work required of those who lived in the capital.

In the early modern period, moral and scientific conceptions of climate found expression in the emergence of climatic botany, which found its most vocal advocates in the wet tropics, regions that botanists for over half a millennium have seen as synonymous with dense vegetation and rain.[26] Christopher Columbus (1451–1506) theorized that the luxuriant foliage of tropical plants in the Americas caused the intense afternoon downpours that he observed on his visits to what are now the Bahamas, Cuba, Haiti and Hispaniola.[27] Naturalists held diverse views about whether forests influenced rain in temperate regions such as Europe, but there was a general consensus that wet tropical rainforests increased precipitation. For the following centuries, many European travellers observed a connection between deforestation and climatic change following in the wake of European trade and colonialism. Nowhere did these changes seem to be more obvious than on small islands such as St Helena in the south Atlantic and Mauritius in the Indian Ocean, where colonial officials and settlers in the 1600s and 1700s destroyed most indigenous forests for fuel and to make way for plantations. Calls to protect forests did not stop deforestation until the mid-1800s, when the conservation movement adopted climatic botany as one of its core beliefs.

Climatic botany was just one of many widely debated ideas in the eighteenth century about how forests influence climate. Although Poivre and many colonial officials on small tropical islands believed that deforestation was bad, many European settlers in North America imagined that the dense forests they encountered made the temperatures more extreme in summer and winter. Chapter One of this book focuses on an international climate debate that started during the American War of Independence, when colonial subjects sought to assert their political and cultural independence from British governance and European perceptions of North America. The chapter explores the colourful debate pursued between Thomas Jefferson, the American republican statesman and future president, and the Frenchman Georges-Louis Leclerc, then Europe's most celebrated naturalist, over questions about the history and future of North America's climate. Leclerc, otherwise known as Comte de Buffon, saw the Americas as having a harsher climate that produced smaller mammals and created worse conditions for sustaining agriculture and European society. Buffon imagined that the extensive forests of the Americas created more extreme hot and cold seasons, led to higher humidity and created large weather events, such as massive thunderstorms and blizzards. Jefferson and other American intellectuals believed the climate of the Americas could be, and was being, improved because of deforestation. Chapter One provides a wider intellectual history of ideas of climate in the early modern period that contextualizes later historical developments.

Attitudes towards deforestation in Europe and the Americas darkened in the early 1800s as climatic botany gained new champions armed with recent findings. Climatic botany required a popularizer to make its way into policies throughout Europe and European empires. Chapter Two explores how the writings of Alexander von Humboldt (1769–1859), a Prussian polymath, helped to popularize the narrative of human-induced climate change to the European scientific community around the world. His views originated from his scientific itinerary across South America during the Napoleonic Wars. Humboldt's first-hand observations in what is today Venezuela prompted him to warn that the destruction of forests caused a collapse in rainfall that devastated agriculture and water levels. His ideas reverberated across both sides of the Atlantic throughout the entire nineteenth century. Humboldt helped handpick a French protégé to revisit South America to follow up on his observations to see if the climate had continued to change. This climate debate occurred at a time of growing pessimism about the human impact on nature marked by the emergence of Malthusian thinking, fears of peak coal and the nascent recognition that humans could cause extinctions.

Humboldt's ideas helped reshape forest management throughout the world after he asked his fellow naturalists to lobby government officials to warn about the threats posed by deforestation-induced climate change. Humboldt asked the botanist Joseph Hooker to convince Lord Dalhousie, then the powerful governor general of the East India Company, to stop deforestation-induced climate change from causing ecological and climatic decline in India. The third chapter explores how Hooker's relationship with Dalhousie convinced the latter to draft legislation in 1855 that started a chain of events that saw more than one-fifth of the Indian subcontinent's forests reserved as state property by the end of the nineteenth century. Colonial Indian forestry reforms caused significant upheaval among the Indian peasantry and landowners, who lost access to and control of valuable grazing, forest products and timber. Despite these social upheavals, the British Indian example helped forestry laws to spread throughout the world.

From the outset of forestry reform, critics pushed forcefully against large-scale state takeovers of forest lands that gave states increased power over private landowners. Nowhere did these issues enter public life more than in the United States, where climate change became the focus of a major debate in Congress during Theodore Roosevelt's (1858–1919) presidency. Roosevelt used forestry laws and conservation legislation aggressively to expand federal power, but his perceived intrusion into the rights of states rankled many Senators and Representatives in the American West. Critics of Roosevelt's legislative proposals pointed to scientific uncertainty around the idea that forests increased rain. Chapter Four traces how these political conflicts occurred at the same time as meteorological studies and climatological research on rainfall and forests were emerging as authoritative disciplines. By the onset of the First World War, the idea that forests influenced precipitation lost favour among many foresters in the United States and Europe. State foresters stopped saying that forests increased precipitation or stabilized climate, but they continued to maintain the position that forests stopped soil erosion and increased water in catchments.

European colonialists fantasized about changing deserts into fertile lands covered with farms, forests and lakes. Chapter Five explores the emergence of climatic engineering, a fringe scientific topic that made its way into popular newspapers and fantasy novels by authors such as Jules Verne. One of the main tenets of climatic botany – that transpired and evaporated water created local precipitation – led people to envision evaporation-based scientific concepts to increase precipitation. From the late nineteenth to the mid-twentieth century, colonial scientists proposed several grandiose schemes that sought to flood African deserts, from the Kalahari to the Sahara, to increase rainfall and recreate imagined ancient climatic conditions. Other scientists proposed vast tree-planting schemes to stop desert encroachment and remake what they believed to be a lost forest that once covered the vast, arid Sahel, south of the Sahara.

Climatic botany reached its nadir in the decades after the end of the Second World War. Chapter Six explores how meteorological research on the atmosphere from the beginning of the 1900s to the 1960s increasingly downplayed the role that vegetation played in recycling precipitation at local and regional scales. The question of how moisture travelled through the atmosphere beguiled researchers until the advent of aviation and new technology allowed for the measurement of molecular mass. Several questions could be answered with greater certainty. Was local precipitation caused by evaporation from nearby forests or lakes or did it come from the ocean or far-away places? Was moisture availability in the atmosphere the determining factor in causing rain? The chapter explores how hydrologists and meteorologists investigated and answered these questions from the late nineteenth century until the 1970s, the decade when most foresters, meteorologists and hydrologists believed that evaporation and transpiration, either from vegetation or large inland bodies of water, played only a small role in precipitation.

A new era of climatic botany started in the 1970s when research on the Amazon rainforest in Brazil raised the tantalizing possibility that forests could produce upwards of 50 per cent of the rain in a region. Chapter Seven explores how innovations in climate modelling and the measurement of the movement of water molecules through the atmosphere provided more conclusive evidence proving the large-scale atmospheric recycling of moisture. Today, a growing number of scientists argue that vegetation plays a significant role in regulating local, regional and even continental climates. A new era of climatic botany is upon us.

The revival of climatic botany via supply-side thinking can and should be understood from historical perspectives. The book's conclusion asks readers to understand the intellectual pitfalls and paradoxes that make it difficult to infer simple lessons from the past. Despite these challenges, history, when interpreted carefully and with proper context, provides a variety of insights, which are needed to make effective decisions about forests and climate in the coming century. This book, we hope, offers the beginning of a long and fruitful conversation between historians, scientists and the public relating to forests and climate.

1

REDEEMING THE NEW WORLD

Most people do not think of climate change when they think of the American War of Independence, but ideas about the history and future of North America's climate played an important role in cultural and scientific debates on the future of the fledgling country. In 1781 the British army seized Thomas Jefferson's (1743–1826) Monticello home in Virginia, forcing him to flee and leave behind his job as the governor of the rebellious colony. Rather than using his time to plot how to defeat his enemies, the British, Jefferson instead decided – curiously – to fight his ally, the French. Jefferson drew his quill pen to write *Notes on the State of Virginia* in 1781–2 at the request of the French delegate to the Continental Congress in Philadelphia. Along with a detailed discussion of Virginia's economy and natural history, the book challenged some of the scientific ideas of France's most celebrated naturalist, Georges-Louis Leclerc (1707–1788), more popularly known as Comte de Buffon. Reading *Notes on the State of Virginia* today offers a fascinating window into early modern ideas about human-induced climate change.[1]

Jefferson bristled at Buffon's negative view of North America. Buffon believed that the plants and animals of the Americas were inferior compared to their European, African and Asian counterparts because of the climatic differences of their habitats. Buffon attributed the extreme fluctuations in weather in North America to the continent's massive forests.[2] This argument implied that the American climate was not ideal for European people in terms of health or even morality.

Jefferson, like many other American intellectuals, took offence at criticisms of their newly declared nation. It became fashionable among American intellectuals to talk about forests and climate in America in the decades after the war ended. Many writers, including Jefferson, concluded that the American climate continued to moderate in terms of extreme cold in winter and heat in summer to a large extent because of deforestation. Many early republic American intellectuals saw deforestation and climate change as a good thing.

Jefferson invoked a variety of other evidence to prove the vigour of America and so downplay Buffon's criticisms of it. He believed that the discovery of giant bones of an unknown animal (now known to be a mastodon) in a tar pit in the Ohio Valley proved that huge animals once lived, and still could be living, in America. He remained hopeful that the mammoth might still be prowling around the remote parts of North America. In *Notes on the State of Virginia*, he wrote that the 'tradition' of the Indigenous Delaware tribes is 'that he [the mammoth] was carnivorous, and still exists in the northern parts of America'.[3] When Jefferson commissioned Meriwether Lewis and William Clark to explore the newly purchased Louisiana Territory in 1803 he told the French naturalist Bernard Lacépède in a letter, 'It is not improbable that this voyage of discovery will procure further information of the Mammoth.'[4]

Mammoths came into the debate because Buffon believed that American fauna were smaller than those in Europe because of America's more extreme climate, with hot, humid summers and cold winters. After all, when Europeans came to America, they found no massive elephants or mighty lions and tigers living there. To Jefferson, the mammoth could be a game-changer. It provided an intellectual checkmate to Buffon by proving the existence of a carnivorous predator larger than any big cat in Africa or Eurasia. Jefferson hoped to find a living specimen in the remote parts of America to truly drive the point home.

Jefferson also believed that America's climate was improving, possibly because of the extensive destruction of native forests that happened in the wake of colonial expansion in the original thirteen American colonies. He wrote, 'A change in our climate however is taking place very sensibly. Both heats and colds are become much more moderate within the memory even of the middle-aged.'[5] Jefferson's writing suggests that he agreed with Buffon's idea that forests influenced climate. Many early American thinkers imagined that humans were cutting down the forests fast enough to modify the climate whereas Buffon did not. Jefferson's enlightenment optimism gave him a 'glass half-full' outlook on America's future.

Jefferson's riposte to Buffon formed the central part of a transatlantic debate about the relationship between human activity and climate change. Two issues dominated this debate. The first issue focused on the historic climate of North America and how climate shaped the continent's natural history and people. European intellectuals questioned why North America's climate differed so dramatically from that of Europe and some condescendingly saw the American climate as producing inferior animals and people. Jefferson and other Americans disagreed. The second issue focused on whether humans could modify the existing climate to encourage European settlement. Many American intellectuals agreed that America's climate was harsher than Europe's, but they tried to prove that deforestation was leading to an amelioration of the climate characterized by warmer winters and cooler summers.

These debates were not merely academic. The claims European scholars made about the inferiority of the Americas weighed heavily on the minds of the American intelligentsia in the fledgling republic in the 1780s. The fact that Jefferson appropriated the time to write such a long rebuttal to Buffon – the only book he ever penned – shows the seriousness of his reaction to the accusations. Jefferson absolutely had to find a way of disproving Buffon's beliefs about the Americas to prove the viability of the new republic.

Jefferson had his task cut out for him. He had to do battle against one of the world's heavyweight intellectuals. Buffon's extensive work on Earth's natural history, *Histoire naturelle*, towered – quite literally if you stacked up all 36 volumes – over any other natural history published in the second half of the eighteenth century. As the superintendent of the French king's private garden, Buffon received access to the nation's libraries, gardens and museums. He also commanded immense respect for the breadth of his knowledge of natural history and he was recognized as being one of the foremost naturalists in the world.

Histoire naturelle evolved throughout Buffon's lifetime to encompass almost every domain of natural history. By the time Jefferson dealt with him, Buffon was an old man (he turned seventy in 1777) who had established himself as the world authority on natural history. By the 1770s, the various volumes of *Histoire naturelle* grandly laid out a natural history of Earth and all of life: starting with the origins of the planet in his first volume, published in 1749, and ending in 1788 with the final publication. Throughout numerous successive volumes published across the second half of the eighteenth century, Buffon speculated that the world's climate had radically changed across time. He saw climate change as the great driving force in the geographical distribution of animals and believed that humans strongly shaped the climate. The most recent climatic epic he called the 'epic of man' because humankind's modification of forests, he argued, led to the moderating of climate across all of Europe.[6] Buffon's ideas shaped naturalistic debates, especially about human-induced climate change, throughout the late eighteenth and early nineteenth centuries.

Enlightenment Ideas of Progress and Nature: From Optimism to Pessimism

Buffon's and Jefferson's concerns about the climate of North America revealed wider anxieties about the nature of human progress and the sustainability of modern intellectual and economic development in the late eighteenth century. A long, dark shadow cast by the pre-industrial world threatened at times to engulf whatever progress humans had achieved. The overwhelming majority of the people in the world remained tied to the land.[7] Only a few small pockets of the United Kingdom had experienced the stirrings of industrial revolution, but even there in only a handful of areas. Amazingly, human life expectancy prior to 1800 had not changed significantly since the origins of agriculture some 11–12,000 years earlier.[8]

European intellectuals since the 1600s had grappled with the profound question of whether humans had progressed in key fields of activity compared with the ancient Greeks and Romans, the great civilizations of the past. This debate was called the battle of the ancients versus the moderns. Early modern European intellectuals worried about the quality and durability of achievements in science, technology, politics, philosophy and literature. They celebrated advances, but worried that society might fall into decline like the Roman Empire had done. What made these concerns all the greater was the trend towards secular histories and worldviews – the rise of secularism downplayed the importance of the afterlife, thereby making material advancement in this world as a way to alleviate earthly suffering all the more important. Worryingly, no one knew with certainty how much human economic and material development could advance.

Historians now recognize how insecure and small the gains of scientific knowledge and technological improvement were in the early modern era. Though it overshoots the mark to suggest that there was no Scientific Revolution at all, the scattered advancements that many celebrate today – Galileo Galilei's (1564–1642) sighting of Jupiter's moons, René Descartes' (1596–1650) mechanical philosophy, Isaac Newton's (1643–1727) explanation of gravity – did very little to change the physical world that most people lived in.[9] Agricultural output and population were expanding, but this was caused partly by forcibly removing people from land, a process known as enclosure in Britain, and partly because the population of Europe and Asia continued to grow following its massive decline during the Black Death. The Enlightenment, though profound and important, reflected the thoughts of a few thousand people located in a handful of European cities. Most attempts to improve the economy focused on boosting production in agriculture and increasing efficiency in existing forms of labour.[10] In an important study of Scotland, the historian Fredrik Albritton Jonsson has shown how these schemes began with bursts of extreme optimism during the mid-eighteenth century, but by the end of the century, enthusiasm had morphed into pessimism.[11]

Human progress was only achievable if the Earth had the same fertility and hospitable climate in the present as it had in the past. Classical and medieval ideas of nature raised the possibility that Earth itself had a life cycle, one which began with childhood, led to a fullness of maturity, then slowly declined into infirmity and, finally, death. In Christian belief, Eve's disobedience to God caused the banishment of humans from the Garden of Eden, meaning that Earth, and humanity, had declined from an original, pure state to one of depravity. The belief that humans could no longer live in Eden helped Christians make sense of the various ills that befell human society.

New theories in geology and natural history raised the hope that the Earth's fertility was not in decline. Most Enlightenment European thinkers, including Jefferson, optimistically saw much of the world as being governed by a benign, stable and perhaps even progressive God-like actor. Though secularization had begun to make inroads into philosophical and scientific thought, Europeans still used religion to imbue meaning into and to interpret the natural world. An early nineteenth-century metaphor for nature from natural theology suggested that God had ordered the world and allowed it to run like a clock. This metaphor reflected developments in technologies, such as watch-making and the rise of mechanical philosophy by thinkers such as Descartes.

Optimistic thinkers believed that the natural world produced a cornucopia or plenitude of life that was regulated by an orderly, outside force, be it a religious or unknown natural cause. Providentially, nature was self-regulating and did not require human action to run. For most people, this invisible force was the hand of God, but a secular strain also developed. The Scottish philosopher David Hume (1711–1776), himself an avowed atheist, posited some unknown 'active cause' that produces 'an immense profusion of beings, animated and organized, sensible and active . . . impregnated by a great vivifying principle . . .!'[12] Humans obviously did not create nature, and therefore, philosophers reasoned, humans could have little impact on how it functioned. For instance, humans might be the catalyst for change on a local level, but they wouldn't be able to cause universal change: people could eradicate wolves from a specific region or country, but few imagined that wolves could be eradicated *everywhere* in the world. The idea that animals could become extinct, either through natural means or human action, was proposed only towards the end of the Enlightenment.

Within this early modern worldview, many European naturalists liked to imagine that the world was the playground of humans, especially men (who comprised the vast bulk of naturalists). Men sat at the top of the 'Great Chain of Being', a hierarchy of all life cascading down from man to woman, to all the other animals below. Parrots, amusingly, were often placed just below humans, because they, of all animals, could talk.[13] This decidedly human-centric attitude drew on passages from the book of Genesis in the Old Testament that emphasized how God created the Earth for man. Christian Europeans imagined their place in nature differently compared with members of many of the world's other major religions, such as Buddhism, which denied the material world's existence, or Hinduism, which made an animal, Brahman cattle, sacred.[14] Historians have argued back and forth about the long-term implications of this anthropocentric attitude. In 1967 Lynn White Jr famously argued that the Christian interpretation of the Old Testament caused many of the world's ecological crises because Christians saw nature as God's gift to humans to use as they saw fit.[15] Others point out that Christians and Christian thought, in religious and secularized forms, helped to launch modern conservation and environmentalism.[16] The reality is that Christianity provided the justification for both the destruction and conservation of nature at different times and in different places.

Not every Enlightenment thinker maintained a benign worldview. The radical editor of the great French *Encyclopédie*, Denis Diderot (1713–1784), and his wealthy friend Baron D'Holbach (1723–1789), challenged the idea that there was an underlying natural order to further undermine France's *ancien régime.*[17] Pro-establishment thinkers also raised troubling questions about the nature of climate change and the fixity of species. Naturalists began to grapple with the thought that the Earth was much older than anyone had imagined, and that it had changed forms drastically over time. James Hutton (1726–1797), a Scottish geologist, argued that the world had an ancient geological history and that the Earth's past climate and geology was strikingly different from the present. Enlightenment naturalists tended to perceive the world as changing gradually, however, rather than rapidly. For instance, Buffon concluded that Earth was warmer in the past, but had continued to cool and become drier as different epochs progressed. To explain fossils, Buffon concluded that many animals became smaller as Earth's climate cooled, because the world's climate lost the ability to sustain such large creatures. This explained why mammoth bones were bigger than those of modern elephants. Buffon's argument disturbed European colonists in the New World and at least one leading nineteenth-century European naturalist, Alexander von Humboldt, because Buffon's ideas suggested that the Americas could never reach the same level of economic, social and agricultural development as Europe.[18]

Historian Mark Barrow notes that the idea of extinction was 'anathema' to almost every Enlightenment naturalist because it implied that Nature could be flawed, thus suggesting the blasphemous thought that God created imperfect creatures and an imperfect world.[19] Some thinkers attempted to resolve the problem of fossil evidence by offering up the possibility that God's creatures were constantly becoming better and better – a sign of His benevolence. Erasmus Darwin (1731–1802), grandfather of Charles Darwin (1809–1882) and an influential thinker in his own right, advanced his own views on the elasticity and abundance of nature. He asked the question, '[are] all the productions of nature . . . in their progress to greater perfection?'[20] Georges Cuvier (1769–1832), the French naturalist who developed an early conception of extinction, took a different approach to the same problem by arguing that extinctions had occurred in the past, but that God always created new species for each successive generation in order to make them perfectly adapted to their new environments – each creation was unique and perfect, without the intermediate forms or monstrosities required by more materialist conceptions of evolution.[21]

The question of progress and nature's fecundity varied in importance for Enlightenment thinkers depending on their economic, geographic and social situation. Buffon's prognostications about the world were non-threatening to him personally – he was ensconced in the highest society in Paris and expressed few concerns about the fate of his own country. Yet to settlers in the New World, predictions by Buffon took on added importance because they raised the possibility that their efforts at colonization could come to naught. Understanding the limits of nature and progress also meant more to advocates of the Scottish Enlightenment because they measured improvement in increased agricultural yields. The failure to improve nature in the Highlands of Scotland raised new, disturbing questions about the limitations of human population. Thomas

Malthus's (1766–1834) dire calculation that human population growth would outstrip the growth in food supply reflected a new type of pessimism that ended the Enlightenment. The possibility that Enlightenment thinkers took an overly optimistic view of the world remained a very real possibility throughout the entire second half of the eighteenth century.

Questioning the Americas

Enlightenment philosophers recognized that European settlement in the Americas offered an opportunity to understand not only the rules governing human societies, but how climate affected governance. Though some philosophers idealized the New World and its 'noble' Indigenous inhabitants, Buffon saw it as an inferior place compared to Europe.[22] In Buffon's mind, the animals in the Americas and Eurasia shared a common origin. This was a radical proposition for the time because quite obviously the animals in America and Europe looked different. Suggesting that two different-looking animals shared a common ancestry meant that in the distant past these animals had changed form. He theorized, drawing from the bones of 'mammoth' animals dug up in Siberia and North America, that the climates of much of the world had changed in the past. For example, he believed that gigantic elephants once roamed a much-warmer Siberia. The bones found there revealed such an animal. But as the climate changed, these animals migrated southwards and became smaller.

Buffon proposed that the crust of the Americas had been submerged by water and arose at a far later date than did Europe. Hence, the Americas had not fully 'dried out' yet whereas Europe had. The extensive forests compounded the prevailing dampness to make the weather more extreme than that in Europe. Buffon believed that Europeans had moderated their own climate by cutting down the once-extensive forests whereas the forests in much of North America made it hot and humid in summer and bitterly cold and snowy in winter. In short, forests created extremes.

The idea that extensive forests increased rainfall gained ground in the eighteenth century. The strongest champions of the idea that forests influenced climate were doctors who dabbled in botany – at the time required study for those training in medicine – and ancient history, a subject that posed the question of how climate had changed in places such as the Holy Land. John Woodward (1665–1728), a physician in London, commented in 1708 that nations with extensive forests had 'great Humidity in the Air, and more frequent Rains, than others that are more open and free'.[23] This made anecdotal sense given that the Mediterranean had fewer trees and less humidity and rain. Northern Europe tended to be more forested and to have more regular rainfall. This theory suggested that as deforestation occurred in northern Europe, the climate would change.

The historian Edward Gibbon (1737–1794), author of the monumental *The History of the Decline and Fall of the Roman Empire* (1776–89), noted that many scholars 'have suspected that Europe was much colder formerly than it is at present', before adding that 'the most ancient descriptions of the climate of Germany tend exceedingly to confirm their theory.'[24] He pointed out that the Danube and Rhine consistently froze over during the time of the early Roman Empire, something which did not occur in Gibbon's lifetime. These findings suggested that Europe had warmed considerably since the Roman period.

Deeply ingrained cultural values influenced how Europeans viewed forests. Most Enlightenment philosophers believed that civilization relied on agriculture, and they tended to live in places such as France or England where many of the wildest forests had been tamed. People who lived in forests were seen as more primitive and barbaric than those who farmed, herded or lived urban lives. The French philosopher Charles Montesquieu (1689–1755) advocated the cutting of forests because he believed that forest clearance improved nature and climate. The Scottish Enlightenment philosopher David Hume and the German philosopher Immanuel Kant (1724–1804) agreed with Montesquieu, believing that forest clearance and agriculture had raised the temperature of once-cold Europe into moderation, with broad fields and meadows replacing cold, dank undergrowth and swamps.[25]

To the early modern mind, forests symbolized dangerous places. This fear is seen clearly in literature from the period. In most traditional variations of the fairy tale 'Little Red Riding Hood', Little Red Riding Hood goes into the forest and never comes out – she and her grandmother are eaten by a wolf.[26] With a sadistic twist, the 1697 version of the tale by Charles Perrault (1628–1703) tells how the girl hides herself from woodcutters who would have been able to save her from death because she does not want to be turned back from her journey. These fears of primeval forests full of danger were not entirely unjustified. Wolf attacks became increasingly common in France when Perrault wrote his story. In 1719 Daniel Defoe's *Robinson Crusoe* described hundreds of ravenous wolves hunting people for food in the forests during winter.

It is in many ways a wonder how such dense forests existed in the eighteenth century given the extensiveness of deforestation in Europe over the preceding two centuries. For thousands of years Europeans had cut down the native forests that once covered much of the continent and Britain to make fields and farms and to provide fuel for homesteads.[27] In the early modern period and before, Europeans believed that the climate reflected not only physical conditions, such as the latitude or the intensity of the sun, but the moral, economic and political conditions of the people who inhabited a region. Europeans by and large saw extensive forests as sites of vice and sources of disease and perceived them as evidence of humanity's failure to domesticate nature, as prescribed in Genesis. The giant American forests were seen as a sign of the failure of settlers to conquer nature and as an environmental and moral impediment to further growth. North America's early European settlers feared large forests and sought to convert them into agricultural lands, but many worried that the process was not happening fast enough to modify the climate or morals of the people.[28]

Buffon's pessimism was not entirely surprising given the early history of British settlement in North America. Europeans, especially the first British settlers, had long complained bitterly about the cold American winters. Of the 102 Puritans who settled in Plymouth, Massachusetts, 45 of them, almost half of the settlement, died during the first winter. Unlike the damp but more moderate English winters, American winters produced heavy snows, below-freezing temperatures and gale winds that could blow in from the Atlantic or down from the Arctic. The memory of the Pilgrims' first miserable, deadly winter weighed heavily on the minds of colonists in the decades after. Advertisements in Britain painted a rosy picture of the climate to induce investment and settlement, but the reality on the ground was that America's climate differed vastly compared to Europe.

America's weather puzzled early seventeenth-century English settlers, who had expected to find warmer climes there than they did.[29] They imagined the American South as having a tropical climate capable of growing tropical produce and Virginia as having an ideal climate to grow wine and olives. However, Virginia, though hot and humid in the summer, suffered through colder winters than England. The colonists compared these extremes with the more moderate winters and summers in southern Spain or Italy, which occupied comparable latitudes in Europe. In the hot parts of Europe, such as the Mediterranean, rain fell in the winter but almost never in the summer. Along the eastern seaboard of North America, rain fell year-round, and rainfall was particularly heavy in the summer. By the middle of the seventeenth century, the paradoxical reality became apparent: most of America had hotter summers than southern Europe and colder winters than northern Europe. This disappointed colonial advocates in England who wanted to cultivate Mediterranean and tropical plants to break the power of Spain and Portugal, then dominant imperial powers in the Atlantic and Indian Oceans.

Over time the British settlers learned to tolerate the cold. The new colonists hewed tall white pines to build solid wooden houses and to fuel roaring hearths.[30] They stockpiled grains, drawing upon farming techniques from Indigenous Americans. In Virginia, settlers planted tobacco for export to the growing number of London coffee shops and their patrons of the coffee-houses who started to smoke pipes. The settlers slowly prospered by becoming traders and farmers who sold primary materials to the Caribbean and Britain. They overcame the obstacles of winter and the challenges posed by a different environment.

But despite the progress of the early colonists, certain fears still lingered in the minds of European intellectuals who watched from across the Atlantic. Would the coldness of the winters and the heat and humidity of the summers change Europeans into different people, making them more like America's Indigenous inhabitants and less like their relatives across the Atlantic? At first, the English themselves feared this. Many writers described the personality and political attributes of the English people as being a product of that country's mild, temperate climate. Edward Hayes warned his fellow Englishmen about the dangers of seeking to colonize the 'hoatt and untemperate Regions' of America, because it would bring 'unto our Complexions intemperat and Contagiouse. Nature hath framed the Sparnyards apt to suche places. Who prosper in drye and burning habitations. But in us she abhorreth suche.'[31] This theme recurred throughout the history of British settlement around the world, first in America and later in the Indian subcontinent, Australia and southern Africa.

Buffon put the idea of degeneracy on firmer ground than other writers, though he was not the only one to make this argument. He pointed to the mammoth as an example of how climate change could lead to degeneration. Clearly a larger elephant-like creature lived in Siberia in a previous era. The discovery of large bones was irrefutable evidence. Like naturalists before him, Buffon did not think that animals went extinct. Rather, he reasoned that the changing climate forced species to transform – in this case, mammoths migrated south to their current home in Asia and Africa, where they shrank, or degenerated, because of their changed environment. Climate change and migration could lead to degeneration.

Even more troubling to American settlers, the idea of the degeneracy of animals applied to humans. People themselves could change, and for the worse. Drawing from Buffon and other European observations of the Americas, Cornelius de Pauw, in his 1771 publication, *Recherches philosophiques sur les Americains* (Philosophical research on the Americans), argued that the environmental conditions of North America had led to the degeneracy in the Indigenous people who lived there.[32] He argued that the vast forests produced unhealthy vapours which harmed human development.[33] Though the American settlers pointed to their progress, the fears brought up by Buffon and nailed down by Pauw raised worrying questions about the future settlement of America. How long could the fledgling nation prosper in this climate?

Saving the New World

Thomas Jefferson felt the sting of Pauw's and Buffon's sweeping arguments and saw them as accusations against America's colonial project itself. A stain of illegitimacy would hang over the heads of the leaders of the rebel American colonies unless Buffon was refuted. If Jefferson himself could not prove the vigour of the New World, perhaps this meant that Buffon was correct? Trying to prove this – and to show that the settlers had modified and continued to modify the climate – ended up being one of the looming intellectual issues in the last quarter of the eighteenth century. This was a time of tension that threatened to tear apart the social fabric of colonial America. From 1775 to 1783 a ragtag group of American revolutionaries fought against loyalists and the powerful and well-financed British army and navy. Jefferson plotted a real war while also battling against Buffon with his quill.

In 1780 the opportunity to refute Buffon and legitimize the American rebellion fell into Jefferson's lap. The French government in Paris, eager to gather intelligence on the capabilities of the American revolutionaries, asked its representative in Philadelphia, the Marquis of Barbé-Marbois (1745–1837), to find out more information about America. In 1780 the marquis circulated a questionnaire to each of the representatives of thirteen states at the Continental Congress.

The questionnaire asked about each state's geography, economy and history. Jefferson received a copy and was asked to answer it for Virginia. He seized his chance to write a defence of Virginia in 1781.

Without the job of governing Virginia, Jefferson took to the task of completing the survey. Jefferson worked to set the record straight. Had the British military conquest of his home at Monticello not taken Jefferson's attention from governing Virginia, it is possible that Buffon would have won the debate. But Jefferson had the time now to spend on crafting a riposte and he used it decisively: he compiled the most up-to-date and comprehensive survey of America's natural history in a single volume, *Notes on the State of Virginia*, which he completed in late 1781 and revised in 1782.

Jefferson divided the chapters in the work by the 22 queries given to him by Barbé-Marbois. The chapter headings focused on the practical questions of topics such as 'Geography', 'Climate', 'Mountains', 'Laws', 'Religion' and 'Aborigines'. Jefferson detailed the geography and natural history of not only what is today Virginia – which he did with unerring accuracy for the day – but the areas west of Virginia that the state laid claim to. The book extolled the strength of the American colonies and refuted Buffon's arguments.

In his *Notes*, Jefferson sharply criticized Buffon's notions that the quadruped mammals in the Americas paled in comparison with their heartier European and Asian counterparts. He disagreed with Buffon that all animals in North America derived from Euro-Asian ancestry, and chided the Frenchman by writing:

> When the Creator has therefore separated their nature as far as the extent of the scale of animal life allowed to this planet would permit, it seems perverse to declare it the same, from a partial resemblance of their tusks and bones. But to whatever

> animal we ascribe these remains, it is certain such a one has existed in America, and that it has been the largest of all terrestrial beings.[34]

This argument flipped Buffon's argument on its head: rather than being inferior, America had the largest of all animals in the world. Indeed, Jefferson's theory drew upon the belief that the fundamental laws of nature did not differ between continents. He continued:

> As if both sides [America and Europe] were not warmed by the same genial sun; as if a soil of the same chemical composition, was less capable of elaboration into animal nutriment; as if the fruits and grains from that soil and sun, yielded a less rich chyle, gave less extension to the solids and fluids of the body, or produced sooner in the cartilages, membranes and fibres, that rigidity which restrains all further extension and terminates animal growth. The truth is, that a Pigmy and a Patagonian, a Mouse and a Mammoth, derive their dimensions from the same nutritive juices.[35]

Another major challenge Jefferson tackled was the question of America's climate. Buffon believed that America's temperatures fluctuated greatly, with torrid heats and humidity pervading in summer and cold Arctic winds bringing blizzards and freezing temperatures in winter. Jefferson imagined, like Buffon, that human action led to climate change. Yet he saw the extensive felling of the forests along the eastern coast of North America as having a profoundly positive effect on the climate and weather.[36] Most importantly, deforestation moderated the harsh winters.

> Snows are less frequent and less deep … The elderly informs me the Earth used to be covered with snow about three months in every year. The rivers, which then seldom failed to freeze over in the course of the winter, scarcely ever do so now.[37]

Jefferson made his points forcefully in the book, but he did not want his book to be read beyond an elite coterie when he published it privately in 1785. Jefferson had completed the book by late 1781 and revised it again in 1782, but he waited to have it printed until 1785 because of the war and his own fears of how some critics might attack his ideas. Jefferson did not want his book to be too widely circulated because he did not want his discussions of controversial topics, such as race and slavery, to be misunderstood or to bring him too much criticism. The book's anonymous English printing in Paris in 1785 (a French translation came out in 1787) brought his ideas to the public for the first time, although his book probably did not fall into the hands of Buffon until 1786, the year when Buffon met Jefferson.

A fortuitous career opportunity provided Jefferson with the chance to dissuade Buffon of his theories. In May 1785 Jefferson took up the important diplomatic post of the U.S. Minister Plenipotentiary to France, at that time America's closest ally. He replaced the widely respected Benjamin Franklin (1706–1790), who had acted as a key conduit of information and garnered French sympathy from 1779 until Jefferson's appointment. Jefferson could not have had a more perfect opportunity to put his views on America across to French intellectuals in Paris, the centre of European culture, science and intellectual life.

In 1786 Jefferson met Buffon, then an aged man.[38] The invitation happened after the Marquis de Chastellux gave Buffon a copy of

Notes on the State of Virginia. Buffon extended an invitation for Jefferson to dine with him in the world-renowned Jardin du Roi, the King's Garden. Jefferson arrived with the marquis to dine with Buffon. When Jefferson tried to persuade Buffon that he had incorrectly assessed America's climate, Buffon handed Jefferson the most recent volume of his *Histoire naturelle* and suggested he read it for insight. Jefferson had clearly not convinced Buffon of his ideas about the animals and climate of North America, although he found through conversation that Buffon had not heard of or seen many of the larger mammals in North America, such as the elk and moose. They enjoyed their dinner but left the table fundamentally disagreeing with each other about the differences between Eurasia and America.

While this might have put him off, Jefferson took away from the meeting with Buffon an even stronger desire to convince him about the size of animals in North America. Jefferson actively tried to change Buffon's beliefs about the inferiority of America by collecting and sending him specimens of big cats and other animals, such as the moose. This new evidence supplied by Jefferson differed from what Buffon had originally examined. Prior to Jefferson's visit, Buffon received what we now know to be misinformation: the temperature he cited for the 'tropical' parts of Latin America came from readings taken at the top of mountains. No wonder he thought that the whole of the Americas, even the supposed 'tropical' latitudes, were frigid.[39] As for his theory about the cold-bloodedness of Indigenous Americans, he had simply followed a long line of great Enlightenment thinkers in making pronouncements about American peoples' habits without ever having met any; Rousseau never saw a real 'noble savage' and Voltaire never met an Amazonian – because the two philosophers never actually visited the Americas.

At one point, Jefferson even asked the governor of New Hampshire, General John Sullivan, to procure and stuff a moose for him in order to show Buffon how large American animals could truly be.[40] The general sent soldiers into the woods to find and shoot a moose, which they did, and then he sent the moose skin, head and antlers to Paris, where it arrived in October 1787. A very serious matter – the question of the vigour of America itself – had some amusing dimensions: the antlers that arrived in France were not original to the moose (and were smaller); scholars now theorize that the fur would have probably fallen off; and the entire skin of the animal was fetid. But still, Jefferson hoped to get the moose to Buffon, who had left Paris because of sickness. Most likely Buffon never saw the makeshift moose.

The debate about the warming of America intensified after Buffon's death. A number of prominent American intellectuals wrote on the subject. In 1792 the American surgeon William Currie looked forward to when, 'in the course of time, this continent becomes populated, cleared, cultivated, improved . . . the bleak winds become more mild, and the Winters less cold.'[41] Perhaps the most important voice was that of Samuel Williams (1743–1817), professor of mathematics and natural philosophy at Harvard University. He agreed with Jefferson that human activity caused positive changes all along the eastern seaboard. In 1794 he wrote in his influential *History of Vermont* that 'this change can be seen all across New England.' 'The Hudson, once frozen solid regularly, is now less common. The winter season arrives later, and is less severe. This change can happen within two or three years of settlement.'[42] Williams argued that the climate of the United States had not remained 'fixed and stable', but was instead 'perpetually changing and altering'. Even more, the climate change had been 'so rapid and constant' that it had become merely a common observation and not a matter of doubt. While this change of climate was clear across the nation, newly settled territories were particularly patent exhibits for it, argued Williams. When settlers moved into a new frontier, 'their first business [was] to cut down the trees' and sow grain. What impact did these activities have? First, they dried out the soil, making it and the air warmer. 'The whole temperature of the climate [became] more equal, uniform and moderate.' As the ground dried up through increased evaporation, streams disappeared, rivers reduced, run-off from rain became rapid and the sea, not the land, held the rain. With less snow, the settled areas were warmer and more temperate.[43]

Not every American settler agreed that the climate of America was changing. The famous senator and lexicographer Noah Webster believed that the idea of climate change was a fallacy built on the misreading of classic literature and misguided anecdotal observation. Webster countered in 1799 that the claims of a warming North America were based on 'a very slight foundation'.[44] He delighted in pointing out the contradictions of famous authors by pointing to other passages in classical writings that showed how climate in the past was like that in the present. He pointed out the inconsistencies of Jefferson's and Williams's descriptions about how America was warming by showing the contradictions in their theories. 'Mr. Jefferson supposes a diminution of the heat of summer. Dr. Williams supposes a general increase of heat in our climate; and I leave them to adjust the difference between themselves.'[45] The debate about whether or how forests shaped the climate remained an unresolved scientific and political debate well into the nineteenth and twentieth centuries, as discussed in the next chapters.

2

WHEN CLIMATE CHANGE BECAME BAD

For much of the twentieth century, generations of American children learned in school that a squirrel in pre-colonial America could run from the Atlantic Ocean on the east coast all the way west to the Mississippi river without touching the ground. Though inaccurate, this anecdote roughly conveyed the sheer scale of deforestation that occurred in the Americas during the eighteenth and nineteenth centuries. North America alone lost forest cover roughly the size of California, 45 million hectares (111 million ac) in total, from 1700 to 1850.[1] Towering trees came crashing down or they dotted the landscape as charred sentinels, burned by fire to make way for farms. Colonists in South America similarly pushed back the green blanket of tropical forest. Plantation owners cut down trees to plant sugar, indigo, cotton, coffee and other valuable crops to be exported to entrepôts throughout the world. New economies emerged. In Brazil, an empire of livestock thrived and multiplied on grasses fed by nutrient-rich ash from recently fired forests.[2]

The stark ecological changes wrought by expanding colonial frontiers seemed particularly striking to first-time visitors, who jotted down their observations as they travelled. Nowhere were first-hand observations richer and more shocking than in tropical regions of the American South, the Caribbean, Brazil and Spanish Venezuela, where plantation owners coerced slaves and Indigenous inhabitants to grow cash-crops. Visceral sights, sounds and smells – the crack of whips, bells regulating time, songs, the boiling cauldrons of sugar and the smell of sweat – created strong impressions on visitors. The embers of freshly burned forest, dense thickets of sugar cane and the patchwork of uncut forest next to freshly ploughed farms created a strong impression that spurred reflections on nature, economy, society and politics.

One influential observer of these scenes helped turn the global tide of opinion about deforestation and human-induced climate change. Largely because of his observations, naturalists and leading intellectuals around the world developed a new appreciation for the importance of trees for climatic regulation. From seeing deforestation in the Americas as a potentially beneficial thing, deforestation became unquestionably bad. Alexander von Humboldt, the young, handsome and wealthy naturalist who captured the imagination of Europe and the world through his exploration of South America from 1799 to 1804, revolutionized how scientists conceived of the relationship between climate, geography and vegetation. The results of his voyages were published in thirty volumes, the first in 1805.[3]

Humboldt was possibly the only person in the world whose personal observations could have commanded so much scientific respect and inspired so many. Young Victorian naturalists worshipped him and sought to emulate his explorations. Charles Darwin idolized him, following in his footsteps on his eventful journey to South America, where he made many important observations that ultimately influenced his theory of evolution by natural selection. Humboldt's considerable wealth allowed him to be a patron of leading male scientists and he spent most of his money travelling and publishing his seven-volume *Personal Narrative* (1819–27) and other findings from his American sojourn. Towards the end of his career, he published his five-volume *Kosmos* (1845–62), which offered a unified physical analysis of the world.

Humboldt's scientific reputation was based largely on the observations he made on his journey to South America and North America from 1799 to 1804. He travelled across the Americas collecting a plethora of information that allowed him to make observations on almost everything – temperature, magnetism, altitude, chemical analysis, electricity and vegetation – about the region. He was a rare mind. In one letter of April 1799 he noted, simply and without bragging, 'My single true purpose is to investigate the confluence and interweaving of all physical forces.'[4] What better place to do that than some of the remotest reaches of South America, where he had been granted privileged access by the Spanish government in the name of science. Prior to the French Revolution, the Spanish had largely stopped foreign naturalists from visiting South America so as to protect its valuable resources and knowledge of the continent's geography. Much of the continent remained entirely unexplored by Western scientists. Humboldt gained unparalleled access to one of the most biodiverse places in the world. South America offered, as it does still to scientists and travellers today, a dramatic and varied landscape consisting of mountains, rainforests, deserts, upland grassland plateaus and high-altitude lakes.

Travelling widely gave Humboldt unique insights into the relationship between vegetation, geography and climate. Among his many innovations, he created the isotherm, and also laid down the foundations of biogeography and ecology by tracing how vegetation types changed according to altitude. He situated all of this knowledge in a single, coherent schematic framework – his belief in a 'general equilibrium' of forces.[5]

As a result of his travels, Humboldt concluded that human action had indeed worsened the climate of tropical regions of the Americas. The historian Greg Cushman notes that 'Humboldt's freedom to travel, in contrast, enabled him to see enough of the Americas to become convinced that human activities were changing the climate for the worse on a hemispheric scale.'[6] This wide-ranging experience distinguished him from other travellers in the Americas, who lacked the necessary government approval, money and tools to draw the inferences that Humboldt did. The legacies of his observations were far-reaching. Andrea Wulf in her biography of Humboldt, *The Invention of Nature*, suggests, 'As Humboldt described how humankind was changing the climate, he unwittingly became the father of the environmental movement.'[7]

Humboldt could better be described as more of a surrogate rather than a father of ideas of climate change because he amplified and consolidated eighteenth-century ideas of climatic botany.[8] In many respects, Humboldt was a popularizer of human-induced climate change rather than an original thinker. What set Humboldt apart from earlier observers was that his prestige, wealth and social connections gave his views weight in scientific and popular circles. His ideas on forests and climate were also well timed.

Ideas of climate change had begun to shift in the early nineteenth century. Earlier optimistic assumptions about the human ability to positively change climate had been challenged by various events, such as the outbreak of yellow fever in the 1790s and failures of acclimatization in Scotland, discussed in this chapter. Ever practical, Humboldt not only linked deforestation to climate change, but argued that it created wood shortages, a longstanding concern in wood-scarce Europe.[9] In doing so, he created an argument that justified a century's worth of government appropriation of forests throughout the world.

It is not stretching the point too far to suggest that the world's system of forest reserves – the largest networks of protected areas in the temperate and tropics – were to a large degree inspired and justified by Humboldt.

Humboldt's views on climatic botany helped to settle competing ideas about the relationship between forests and climate. The idea that deforestation could improve and moderate climate still had many advocates in temperate regions of Europe and North America. To the contrary, naturalists and officials on tropical islands saw deforestation as a destroyer of climate. Both perspectives agreed that forests influenced climate, but it was the outcome on which they differed. In 1811 the American statesman and physician Hugh Williamson (1735–1819) sought to resolve this difference by pointing out that climate change could be positive or negative depending on the climate. In *Observations on the Climate in Different Parts of America*, he wrote:

> The cooling process affects every country more or less. The heat would be intolerable in low latitudes, if the process did not exist there, to a great degree. A perpetual verdure and thick foliage, within the tropical regions, tend greatly to moderate the heat . . . In high latitudes, when the country is not mountainous, by exposing a smooth surface, without much timber, to the influence of the sun, the inhabitants may enjoy a temperate climate.[10]

Williamson's theory elegantly explained why people living in the temperate climates of North America should cut down its vast forests, while people living in the tropics should not.

Humboldt resolved this paradox by emphasizing the overall negative consequences of deforestation for *all* climates. His argument implied that the reckless destruction of forest cover set off a cascade of negative ecological, climatic and economic reactions for all locales. Humboldt laid out the central premise of human-induced climate change that informed most nineteenth-century conservation efforts. To make his case, he pointed to the example of Lake Valencia, in Venezuela, where the water had been steadily receding for decades. European planters had cut down the forests along the edge of the lake to create plantations. He argued:

> The changes which the destruction of forests, the clearing of plains and the cultivation of indigo, have produced within half a century in the quantity of water flowing in on the one hand and on the other the evaporation of the soil and the dryness of the atmosphere, present causes sufficiently powerful to explain the progressive diminution of the lake of Valencia.[11]

He took this point a step further by suggesting that Lake Valencia evidenced a global problem for *all* climates. He wrote, 'By felling the trees which cover the tops and the sides of mountains, men in every climate prepare at once two calamities for future generations; want of fuel and scarcity of water.'[12] The local effects of deforestation caused negative consequences throughout the Americas, in tropical and temperate climes. Cutting down forests lowered streamflow – that is, the volume of water and rate of flow in a stream – in rivers and led to declining water tables. 'When forests are destroyed, as they are everywhere in America by the European planters, with imprudent precipitancy, the springs are entirely dried up, or become less abundant,' Humboldt stated.[13] On his journey through Mexico, Humboldt saw a similar process of falling lake levels at Lake Texcoco.

Not everyone at the time agreed with Humboldt's explanation for the declining waters of Lake Valencia. François de Pons (1751–1812), agent to the French government in Caracas, offered a rival theory that challenged Humboldt's belief that evaporation alone had drained water in the large river-fed lake. Like Humboldt, Pons travelled widely throughout South America. During his journey across the continent in 1801–4, and after Humboldt's own visit, he visited the lake, and relied on the 'concurrent testimony of my own eyes, and that of the intelligent Spaniards who live in the vicinity' when formulating his theory that a subterraneous passage discharged the water.[14] He scoffed at the idea that evaporation could discharge such a large volume of water: 'Would evaporation alone, great as it may be in the tropics, have been adequate to the consumption of so great a quantity as the rivers supply?'[15]

Humboldt's theory gained increased public attention because Pons' views received so much scorn; criticism of Pons brought praise to Humboldt, who published his initial findings on Lake Valencia and Lake Texcoco in 1805, two years before Pons' publication.[16] Upon publication, Pons' theory was quickly ridiculed.[17] In 1819 Humboldt went further out of his way to dismiss Pons' already depreciated view. In the first volume in 1819 of his *Personal Narrative*, he wrote, 'I am not of the opinion of a traveller, who has visited these countries since me, that "to set the mind at rest, and for the honour of science," a subterranean issue must be admitted.'[18] Humboldt's views became the standard explanation in popular travel books and encyclopaedias. In 1825 the writer Josiah Conder declared confidently that Humboldt's views were now 'modern science' and 'evaporation is a cause quite adequate to explain the phenomenon of lakes fed by rivers, yet having no channels to discharge their waters'.[19] The 1824 *Encyclopaedia Britannica* entry on Caracas concluded against

Pons more gently: 'How far this conjecture is well-founded, it seems extremely doubtful.'[20]

The question of the declining water level of Lake Valencia, though settled in Humboldt's favour, still lacked the verification he wanted. If Humboldt could send an emissary back to the lake to observe its levels, then perhaps he could have further empirical proof. What had happened to the lake? Had it continued to decline, as many expected, or not?

IN THE TIME SINCE Humboldt left, much had changed in South America. The French Revolution and Revolutionary Wars in Europe (1793–1815), and the upheavals they caused in the Americas during the 1790s to 1820s, redrew the political boundaries and ideologies of the Americas. The social order that Humboldt saw and lamented – monarchy, aristocracy and slavery – underwent radical change. In response to the French Revolution, in 1791 slaves in the French Caribbean colony of Saint-Domingue rebelled and, eventually, following years of uprising, warfare and invasion by Napoleonic forces, declared a new republic, present-day Haiti, in 1804. Creole and European elites in Spanish possessions were also influenced by the Revolution. In 1808 Napoleon made his brother, Joseph Bonaparte, the king of Spain, a decision that was unpopular in the Spanish colonies, which established Juntas of loyalists. The restoration of the absolutist Fernando VII (1784–1833) to the Spanish throne in 1814 further inflamed local Spanish and creole elites who worried that the foreign Spanish government would threaten their power and prosperity. A series of protracted guerrilla campaigns and wars from 1810 led to the eventual independence of the Spanish colonies of Mexico, Chile, Colombia, Venezuela, Peru, Ecuador and Panama by the

1820s, although official recognition by Spain only occurred in 1836. Brazil became independent from Portugal in 1822.

Elites in the newly established republics looked to Humboldt as a patron figure. He was the acknowledged world expert on the natural history and geography of South America. He was also a friend of Simón Bolívar (1783–1830), an aristocratic Venezuelan who led the fight for the independence of Gran Colombia (including Ecuador, Colombia and Venezuela) and inspired similar movements from as far south as Argentina and as far north as Mexico. Bolívar's deputy, Francisco Antonio Zea (1766–1822), requested Humboldt recommend experts who could help to put the new state of Gran Colombia on firm foundations.

One of Humboldt's recommended scientists was Jean-Baptiste Boussingault (1802–1887), a promising young French chemist who helped 'prove' to science that deforestation indeed caused climate change. Twenty-two years after Humboldt's fateful visit, he confirmed Humboldt's observation in Venezuela while working in South America on behalf of Humboldt and the Gran Colombian government. Boussingault had received offers to work as a mining engineer in Egypt, Guatemala and Chile but instead, in the words of Cushman, he 'jumped at the opportunity to follow directly in Humboldt's footsteps'.[21] Though he officially worked for the Gran Colombian government, Boussingault received advice from Humboldt on sites to revisit to take new observations to compare with old ones. He unsuccessfully tried to climb Mount Chimborazo in Ecuador and he visited Lake Valencia in Venezuela.

On arrival at the lake, Boussingault found a radically changed landscape. The lake's level had visibly risen, covering former cotton fields. What caused this rise? Boussingault agreed with Humboldt's earlier theory, which seemed to be confirmed by political events that had fundamentally transformed the landscape. The same revolutionary upheavals that led to the creation of the Gran Colombian government, which employed Boussingault, caused a decline in the local plantation economy. He wrote,

> On the first cry of independence raised, a great number of slaves found freedom by enlisting under the banners of the new republic; agricultural operations of any extent were abandoned and the forest, which makes such rapid progress in the tropics, had soon raised possession of the surface.[22]

As a result, evaporation decreased, and the lake's level steadily rose. The results could not be clearer. In a wry tone, Boussingault mocked Pons one last time: 'Those who had explained the diminution of the lake by supposing subterraneous canals, now hastened to close them in order to find a cause for the rise in the level of the water.'[23]

Boussingault published his observations in his book *Rural Economy* (1845). Boussingault's confirmation of Humboldt's original thesis helped, in a sense, to prove that forest cover influenced evaporation and rainfall, and showed that increased forest cover led to increased water supply.[24]

The publication of *Rural Economy* confirmed a theory that had become widely believed by naturalists and large segments of the public: the destruction of forests caused considerable negative changes to climate. Climatic botany advocates found plenty of historical and contemporary evidence that confirmed their worst fears. Scholars of the Bible and of the classical world pointed out that regional climates had been changing – for the worse – for thousands of years because of human action. Foresters advocated sparing large forests throughout the world from destruction in order to ameliorate the local and regional climate. Humboldt's theory, and the support given to his idea, led to national governments creating the world's first system of nature reserves. It was his prestige that helped climatic botany advocates start the world's first global movement to stop human-induced climate change.

Growing Fears of Climate Change and the Limits of Growth

Humboldt's fear that deforestation in the Americas would have negative long-term economic, ecological and social consequences became popularized partly because it struck a chord with naturalists and intellectuals who were worried about the limits of human and natural progress. The late eighteenth and early nineteenth centuries saw considerable political and social changes throughout Europe and the Americas. Revolutions saw the king of France beheaded, the nations of Europe occupied by Napoleon and slave riots in the Caribbean and South America. Despite nascent industrialization in Britain and a slowly rising standard of living, growing numbers of people lived in poorly sanitized industrial cities. Some leading minds argued that humanity had reached – or worse, exceeded – its natural limits. These fears recurred throughout the nineteenth century in response to industrialization, globalization and the growth in the human population. In many respects, humans societies have never left these fears behind.

Humboldt's argument about deforestation was embedded within a wider social critique of colonialism in the Americas. In his mind, the destruction of forests, the creation of tropical plantations and the institution of slavery were all interconnected. Tropical plantations produced negative ecological, climatic and social impacts.

Harsh social conditions begat harsh environmental conditions, a point echoed by other abolitionists.[25] Though Humboldt approved of settlement and agriculture, he disapproved of reckless deforestation, something that evidenced human greed and ignorance. Human-induced climate change, like slavery, could only be corrected through proper liberal governance. Consequently, Humboldt advocated for a socially conscious environmentalism.

Humboldt's vision of environmentalism raised challenging questions. Who should be given power to regulate the causes of human-induced climate change, the state or civil society?[26] Which professional groups can define the issue? Scientists? Politicians? Lawyers? Economists? Citizens? Businesses? The broad outline of positions on this issue, from extreme free market advocates to those who see a strong role for government intervention, originated during the late eighteenth and early nineteenth centuries when modern economics emerged. Ideas about economics and climate change have been intimately linked together for over two hundred years.

As discussed, many Enlightenment figures, such as Jefferson, took a reasonably optimistic view of the world. Jefferson's personal views reflected a mix of humanism, science and secularized Christian theological perspectives about the Great Chain of Being, which imagined that God had divinely ordered the world for the use of humans. Many Christians agreed that God made the world orderly. William Paley (1743–1805), who is credited with formulating an early argument for intelligent design (which is still invoked today by many Christians who disagree with evolution), wrote in 1785 that, 'God, when he created the human species, wished their happiness; and made for them the provision which he has made, with that view and for that purpose.'[27] We credit Paley with using the metaphor of a well-timed clock to describe nature, programmed by God. How else could humans explain the laws of nature and the regulation of the heavens and the seasons?

Adam Smith (1723–1790) and other Scottish intellectuals were influenced by a particularly optimistic strain of the Enlightenment, often synonymous with the views of Anthony Ashley Cooper, the third Earl of Shaftesbury (1671–1713), who believed humans could improve nature by transforming less valuable land into more valuable land. Scotland, especially its Highlands and Hebridean islands, was an intense site of experimentation because its sparsely populated, boggy, grassy, tree-less mountains seemed to call for settlement. Smith conceived of the environment, especially the soil, as a source of wealth and abundance that could be improved through good farming practices, but he did not, like other Scots, believe in the ability to dramatically change the environment in places such as the Scottish Highlands.[28] Like many economists of his lifetime and after, Smith did not worry about climate change, because he saw climate as being more or less fixed. The only requirement to capitalizing on this wealth was human ingenuity and a freer market. Smith showed less interest in changing the climate because he saw existing climatic differences as creating the ideal conditions for exchange. He noted that wine could be produced better and more cheaply in France than Britain, so why shouldn't the British trade for the wine by offering products that were more suited, and thus cheaper, to produce under British conditions?

Some Scottish landowners hoped to improve the climate dramatically by draining bogs and planting trees, two activities imagined to be synonymous in the late eighteenth and early nineteenth centuries with 'good' climate change. Much like Thomas Jefferson believed that deforestation was improving the climate of the New World, so too did some Scottish landowners and thinkers posit that drainage and tree planting was ameliorating Scotland's notoriously cold, rainy weather. The Highland Society of Scotland, founded in Edinburgh in 1784, supported landowners and intellectuals who experimented with these practices. William Aiton (1731–1793), a Scottish botanist who served as the first director of the royal botanic gardens at Kew, in southwest London, argued that the creation of peat bogs was the result of deforestation. He explained this by pointing to the presence of buried roots and logs, which he and others theorized had been cut down by the Romans. To 'restore' Scotland to its rightful wooded state, he advocated draining peat and planting trees that thrived in the boggy conditions. In 1811 he wrote, 'Where the deep fen, or dark morass now lies,/ Tall trees may grow, and richest verdure rise.'[29]

The enthusiasm and optimism of the Enlightenment was fundamentally challenged during the turbulence caused by the French Revolution, which set off wars that raged, on and off, from 1793 to 1815 across Europe, the Atlantic, Asia and the Middle East. In the aftermath of these wars, the victorious conservative European governments in Britain, Prussia, Russia and Austria re-established or enlarged the French monarchy, the Spanish monarchy, the Sardinian monarchy, the king of Sicily and the Portuguese monarchy, and then created a monarchy of the Netherlands for good measure. Many people who had enthused about the possibility of democratic government were crushed by the weight of newly revitalized monarchies.

The Anglican clergyman Thomas Malthus proffered a much more dire worldview than the one advanced by Paley and Jefferson. Malthus famously argued that human population growth, unless controlled by human intervention, would outstrip the increase in food production, resulting in chronic poverty and famine. Ever the pessimist, Malthus believed that God created the world that way and despite efforts from humans, it was largely futile to try to stop it. It would be possible to mitigate this effect somewhat by delaying marriages and thus lowering the birth rate. Unlike climatic botanists, Malthus did not think that humans could alter the climate.[30]

Malthus's idea, stripped of its Christian context (though not necessarily of its religious imprint), has had an enduring influence on scientific and environmental thought since its inception. Charles Darwin used Malthus's idea of resource limitations to conceive of how the 'struggle for existence' drives evolution via natural selection. The ghost of Malthus's postulate lingers in debates about the limits of growth. In 1972 the Club of Rome published the book *The Limits to Growth*, which offered the classic neo-Malthusian viewpoint that exponential human population growth threatened all life on Earth, human and non-human. Today one often hears echoes of Malthus in environmental debates about sustainability. We are warned that, in the face of increasing population growth, the day will come when we will no longer be able to keep up with worldwide food demand.

The idea that Malthus posed has proven to be one of the most enduring riddles of our industrial, globalized modern world. Like many great theories, Malthus's analysis described the past reasonably well, but it failed to predict the future. His pessimism captured an element of human history after the Agricultural Revolution some 12,000 years ago. We now know that human populations prior to the Industrial Revolution underwent periodic setbacks because of disease, famine or large-scale war. Malthus's prediction failed to foresee the centuries after his death; he did not envision that the Industrial Revolution would increase wealth, health and people's caloric intake. The human population exponentially increased in the nineteenth and twentieth centuries, but even with this rapid growth, the world suffered relatively few major famines and those that happened had little impact on the global population. India last suffered famine as recently as the 1940s, but the country has still managed to increase its population sevenfold in the last two centuries, from 169 million in 1800 to 1.4 billion in 2023.

Malthus's argument was one of the more dire viewpoints in a larger debate about natural resources and the limits of growth. Malthus wrote in his *Essay on the Principle of Population*, first published in 1798 and revised in six editions (the last in 1826), that

> the great question is now at issue, whether man shall henceforth start forwards with accelerated velocity towards illimitable and hitherto unconceived improvement, or be condemned to a perpetual oscillation between happiness and misery and after every effort remain still at an immeasurable distance from the wished-for goal.[31]

Almost all nineteenth-century concerns about changing climate to some degree revolved around this central question of the limits of natural fecundity and human growth.

Ideas of human-induced climate change came to prominence in the 1820s to 1850s on the back of a bigger intellectual debate about the limits of growth. The historian Fredrik Albritton Jonsson emphasizes that intellectual thought about the environment during the early nineteenth century centred on two key debates. First, the idea of environmental management revolved around 'rival ecologies of commerce', which pitted naturalists, who argued that expertise was required to conserve natural resources, against free-market advocates, who believed that little to no government regulation was required to protect resources because nature was fecund and self-regulating.[32]

Second, key intellectuals and naturalists debated the limitations of human population growth and the use of natural resources. Enlightenment optimists like Jefferson and Smith believed it was possible to continue to improve human knowledge and quality of life. Pessimists like Malthus believed that the underlying laws of nature resigned much of humanity to misery and poverty.

Pessimism led to alarms about the decline of natural productivity, while optimism provided the justification for trying to solve the problem. Fear is in many respects more politically potent than optimism. Problems require solutions. And for that to happen we need people, experts in this case, to identify those solutions. In the nineteenth century, foresters played the role that ecologists or environmental managers do today: they examined public and private forests and made legally binding recommendations about how to conserve them.

3

STOPPING CLIMATE CHANGE IN BRITISH INDIA

In the 1850s, a British autocrat in India instigated the largest effort to regulate climate that the world had ever seen. Few people have ever had more power – and used it – than the Earl of Dalhousie, when he served as governor general of the British East India Company from 1848 to 1856. On landing in India, he surprised people in the capital of Calcutta with his youthful, clean-shaven face, which made him look even younger than his 34 years. Many in Britain and India wondered whether this fresh-faced young man could manage the enormous East India Company, which ruled over 100 million Indians and stretched thousands of miles from Burma in the east to modern-day Afghanistan in the northwest. He soon erased any doubts of his ability to govern the country.

After landing in Calcutta, Dalhousie made the bold decision to put nearly the whole of India under British control. Before he could receive advice from the governing Company Board in London, he became embroiled in a vicious war with the powerful neighbouring Sikh Empire. He won the war and annexed the Punjab in 1849. He then took Pegu (modern-day Myanmar, previously Burma) in the early 1850s. A letter reportedly sent by the Company in London to India in 1852, prior to taking Pegu, implored: 'We want no more territory – we have too much already.'[1] During his rule, he expanded the borders of British India through numerous invasions and annexations of territory, including that of five princely states, which he 'acquired' under the so-called Doctrine of Lapse, which posed that any land without heirs could be taken by the British occupiers.

Dalhousie travelled throughout the country risking his health and safety to see the diverse terrain of the subcontinent. He often journeyed alone, travelling the many miles on horseback, and even sleeping in a tent with little protection in dangerous places known for harbouring bandits who robbed and killed at night.

The trips on horseback gave him a better understanding of India's diverse environments than any governor general of India before him. He suffered in humid, mosquito-ridden jungles, gazed across the vast arid plains in the Punjab and viewed the mighty Himalayas. This travelling created a deep impression about the need to stop the destruction of forests and to create new forests in arid lands. In February 1851 he drafted a minute calling for tree planting in the Punjab because he found the land 'neither adorned by the foliage, which is its natural ornament, nor stocked with the timber requisite for a thousand purposes in the everyday life of the people who dwell in it'.[2]

Dalhousie became convinced from his travels, reading and conversations that India had lost many of its natural forests over the last few centuries and, as a result, had progressively received less rain. Dalhousie issued a memorandum in 1855 giving the government control over vast swathes of India's forests. The memorandum invoked Indian and Burmese legal precedents that gave princes and kings the right to possess and hunt in forests. Now, Company law could ban local people from collecting timber or other products from forests.

Invoking climate change helped justify the Company takeover, which focused selectively at first on the most timber-rich regions. The conquest of Pegu, with its teak-rich forests, helped prompt Dalhousie's decision. Teak (*Tectona grandis*) was one of the most valued timbers in the world because of its dense, fine grain and resistance to insects. Britain's Royal Navy still required such timber supplies for ships and docks, and India's expanding railways needed sturdy wood for railway sleepers to handle the wear-and-tear of heavy locomotives.

Many of Dalhousie's actions were surrounded by controversy from his arrival in India. The future secretary of state for India, Sir Charles Wood, praised Dalhousie's nerve in the face of public criticism, saying to him how he admired his ability to 'carry on [his] affairs in spite of the press. But what a trial of patience and temper it must be.'[3] Wood was not the last to point out the public controversy which Dalhousie's actions caused.

Dalhousie has been credited with instigating environmentalism by creating forest reserves, but has also frequently been criticized as a colonial authoritarian for setting in motion laws that alienated Indians from the forests they had long inhabited and used. Dalhousie's forestry reform set in motion events that changed the lives of millions of people and millions of hectares of India's forest. Not only this, but the precedent set by Indian forestry also encouraged other countries to undertake similar reforms and provided a model for the establishment of forest systems elsewhere in the British Empire and even in the United States.[4] Forestry reforms also occurred simultaneously in the French, German and Dutch empires. By 1900 most countries of the world had passed legislation that made the state the steward of forest resources; created a department devoted to managing forests based along the lines prescribed by professional foresters; and forced traditional forest users to submit to management authority. Forestry policies often challenged Indigenous land management practices, such as slash-and-burn farming. These basic policies have only started to undergo significant change in the past thirty years.

Dalhousie's conversion to the belief in human-induced climate change highlighted a global trend that began in earnest in the mid-1800s. Prior to 1847, Dalhousie cared little for forestry and thought almost nothing about climate change. Yet he was convinced that humans affected climate based on the groundswell of scientific and popular interest in the subject in India as well as elsewhere in the world. A similar pattern of experiences influenced people in North America, South America, North Africa, Australia, New Zealand and South Africa to come to similar conclusions as Dalhousie: deforestation threatened the climatic and economic stability of modern society. Dalhousie, however, had power that almost no other person in the world had: he could almost singlehandedly impose new forestry legislation to protect forests for conserving the climate. The autocratic structure of the East India Company, and the sheer time it took to communicate between London and Calcutta, meant that Dalhousie could impose his personal whim onto the second most populous country on the planet. His reforms led to the creation of millions of hectares of government-run forest reserves that covered one-fifth of India's total area. But first, Dalhousie had to be convinced that climate change was real.

A Passage to India

Dalhousie's conversion to believing in the dangers of human-induced climate change happened partly because of a chance encounter with a well-connected and brilliant English botanist, Joseph Dalton Hooker (1817–1911). When in September 1847 he went to Southampton to board HMS *Sidon* to sail off towards his official posting in Calcutta, he greeted Hooker for the first time. Hooker, the best friend of Charles Darwin, was one of Britain's most prominent botanists. The pair quickly became friends, although Hooker noted after meeting Dalhousie that, 'I find Lord Dalhousie an extremely agreeable and intelligent man in everything but Natural History and Science, of which he has a lamentably low opinion, I fear.'[5] Considering the fact that Hooker was himself a natural historian and scientist, this was no insignificant point. Hooker's diplomacy worked to a great extent because he was able to fit within the clubby, chummy social world of Victorian Britain. Few other scientists received such close attention from administrators, because science was not considered an elite profession. In the colonies, botanists and foresters were viewed along the lines of engineers, veterinarians and other public servants who were seen as much lower on the social order than judges, district officers, military officers and other high-ranking civil officials.

Dalhousie and his wife liked Hooker's personality, his respectable social standing and, as Hooker wrote about Dalhousie, 'he is much pleased at me being busy and especially with my carrying on my Meteorological register three times a day.'[6] Hooker's industriousness pleased the practical Scot, whose last job was working on the powerful Board of Trade to increase and manage Britain's growing global influence in trade and manufacturing. Jeremy Bentham's utilitarian social philosophy, which espoused that governments should pursue policies to ensure 'the greatest good for the greatest number', strongly influenced Dalhousie's political outlook.

Initially, Hooker's friendliness reflected his desire to influence Dalhousie's scientific policies and to use his valuable patronage. Unbeknown to Dalhousie, Hooker was acting his part in an international plot to convince Dalhousie of the dangers of human-induced climate change in India. Chance may have brought the two together on a nearly four-month-long boat ride, but Hooker had been preparing himself for some time to advocate the importance of forest cover as a regulator of climate. Before leaving for India, Hooker had been corresponding with Humboldt, then one of the great doyens of European science, who pressed Hooker to broach the issue of deforestation-induced climate change with East Indian officials. Little did Humboldt realize just how perfect an opportunity would present itself to Hooker during his journey to India.[7] Hooker's beliefs about climate change had been formed by his relationship with Humboldt and the description of rainfall declines in St Helena.[8]

The pair talked frequently during the boat journey towards Alexandria in Egypt, where they crossed the desert on a horse-drawn carriage on their way to the Red Sea port of Suez. Dalhousie cemented their friendship by asking Hooker to stay with him and his wife in their private suite on the official Company steamer sent from India to pick him up at Suez. Hooker enjoyed his newfound social status and the comfort afforded him by Dalhousie's luxurious accommodations, bragging in a letter to his friend Charles Darwin about how his trip to India 'was a most delightful one' with stops at Lisbon, Gibraltar, Malta, Alexandria, Aden, Ceylon, Madras and finally Calcutta.[9]

What Hooker did not tell Darwin, however, was that Dalhousie lacked an interest in seemingly all theoretical, practical and aesthetic dimensions of botany. During their carriage trip to Suez, Hooker tried to interest Dalhousie in the practical uses of the gum of an *Acacia* tree. But after handing the tree branch to Dalhousie, 'he chucked it out of the carriage window.'[10] Nor did Dalhousie care for a beautiful, rare rose that Hooker pointed out.

Dalhousie rejected Hooker's discussions of Egyptian roses and the gum of *Acacia* because his interests were honed towards a much larger and more practical object: managing the vast hotchpotch that comprised the East India Company. Throughout its 245 years in operation, the East India Company conquered, inherited, gained control over and purchased most of the governments that made up the Indian subcontinent. Excluding Burma and Punjab, which Dalhousie annexed, the Company ruled from the Himalayas to the Indian Ocean, from present-day Pakistan to Bangladesh and parts of Myanmar. British politicians saw the task of properly governing India's economy, polities, religions and peoples as the toughest, most important administrative job in the whole of the British Empire.

Though the Company no longer existed purely to make profits, as indeed it sought to do throughout the seventeenth and eighteenth centuries, it nonetheless required a steady and growing stream of revenue and taxes. This is probably why Dalhousie approved of Hooker's meticulous meteorological studies, which had various practical applications, such as helping to determine what crops could be profitably grown in different regions or where Europeans could live without suffering from tropical diseases. Dalhousie hinted to Hooker during the voyage about the possibility of him studying the question of tea planting in Assam during his stay there.[11] Such a side trip could take extra months out of Hooker's planned itinerary, but Dalhousie and other officials were desirous of breaking China's monopoly on tea production. Hooker, though not initially keen to make the trip, had to keep in mind his patron's wishes.

Upon arriving in Calcutta, the capital of the East India Company, Dalhousie entreated Hooker to stay at his palatial neoclassical Government House. Writing to Darwin, Hooker bragged, 'I have three homes in Calcutta,' because in addition to staying with Dalhousie, he had standing invitations to stay at the advocate general's and chief justice's houses.[12] For three years Hooker remained in India, travelling throughout the northeast regions, including mountainous areas of Tibet where no European had travelled before. During this period, Hooker relied heavily upon the patronage of Dalhousie, who arranged travel for Hooker and even saved his life when the Rajah of Sikkim imprisoned him in November 1849 in a mountain pass in Tibet.[13] In return for Dalhousie's aid, Hooker travelled to Assam at Dalhousie's request. He recommended the planting of tea at Darjeeling, the region that eventually became the centre of India's tea industry.

Hooker never forgot Humboldt's request to convince Dalhousie about the need to protect forests for climatic stability. His belief in climatic botany only grew the longer he stayed in South Asia. During his travels up the Ganges, Hooker witnessed the extensive and rapid denudation of virgin forests throughout the hills and plains. The scale and impact of forest cutting awed him. After seeing mile after mile of cutover forests from February to March 1848 he told his friend Darwin,

> The change in climate consequent on cutting down the timber over the several hundred miles I have traversed (& many hundred more which I never shall see) is as may be supposed very considerable, not only is the annual fall of rain being diminished, but in the diminished frequency of the showers &

> especially of thunder & lightning & all but total absence of hurricanes. When the total destruction of the woods shall be effected the crocodile of the highest ponds will I suppose be suppressed; the streams & other waters already having diminished greatly. We are all familiar with the cause & effect above alluded to, but I know of no instance where the phenomenon is displayed over so great an area & in so short a time as this part of India affords.[14]

Hooker relayed this shocking information to Dalhousie, who was in the midst of planning a campaign against the Punjab, the last great adjacent military power not yet incorporated into the East India Company. Dalhousie was too busy in 1848–9 to act on Hooker's information, but in the early 1850s he began to seriously consider the position of India's forests.

The belief that the British needed to educate and guide the people of India happened concomitant with early observations that India's natural wealth might not be endless. During the Napoleonic Wars, the British admiralty worried that the supply of teak, a valuable timber for making ships, was in steep decline along the southwest coast of India. Local Indian merchants overharvested this tree and sold the timber to Arab, Indian and British buyers. James Watson, a trained police officer, appealed to the Madras government to let him regulate the cutting of teak, a request that was initially denied but later accepted in April 1807. With no formal training in forestry, Watson took over the management of state forests and began a zealous effort to stop the cutting of teak.[15]

Watson quickly offended the unprepared residents of Malabar by declaring in the first sentence of his proclamation that the 'Court of Directors' of the East India Company had 'resolved to assume the sovereignty of the forests'.[16] In reality, Watson pushed for this claim and he, not the board of directors of the East India Company, took responsibility for the control of these powers. Watson became something of a rogue conservator working in an ill-defined legal and administrative area. He expanded his powers rapidly and by 1809 had the power to stop the export of teak – granted in the hopes of weakening piracy in the Gulf of Arabia and Persia – and even the cutting of timber on private property. Many private landowners and businesses that suffered from the conservancy felt that their rights were being trampled on. Complaints to the government grew yearly. By 1815 the conservator's power had grown to such an extent that not only did Indigenous people dislike it, but 'it was complained of by all the local authorities, by the judges, the magistrates and the collectors.'[17] Watson's policies, though they protected existing trees, did nothing to actively encourage the planting of new ones, save for allowing for the natural regeneration of teak. A review in Bengal in 1852 noted that there was no evidence on record that Watson planted a single seed or sapling.[18]

After more than a decade of heated complaints, the governor of Madras, Sir Thomas Munro, successfully convinced the Company's board of directors – based upon his minute of November 1822 – to abolish the existing conservancy in favour of a legal system that protected private property rights. Historians have frequently interpreted Munro's request as symbolizing the victory of laissez-faire capitalism over state environmental intervention.[19] Though Munro's request to end the conservancy was indeed based upon a desire for free trade and individual rights, we should remember that these were also the rights of Indian landowners who demanded that they be able to use their land as they, not the imperial British government, saw fit. Instead of acting as the cog of private interest,

Munro reflected the widespread disdain of the hastily conceived and potentially illegal conservancy.[20] He sympathized with Indians and described the actions of Watson in most negative terms: '[the conservator's] chief business is to invade every man's property to harass him in his own fields – in his barn – in his house and in his temple.'[21] Munro did not deny the ability of Indians to take care of their own environment, though he, like many others of his age, imagined that the market would create an incentive to plant trees if too many were cut down.[22]

After Munro's decision, Malabar and other highland tracts of southern and central India experienced an increased rate of deforestation from Indian timber merchants during the 1820s and 1840s. The timber was harvested for the rapid expansion of tea, cinchona and coffee plantations and the growing domestic and international demand for teak.[23] The steadily increasing population in India increased pressure on limited forest resources. Scholars and nineteenth-century foresters have debated whether Munro's decision led to an increased cut-rate, but there is not reliable enough statistical evidence to prove this one way or the other, nor can historians predict what would have happened had the state monopoly on timber remained. What is clear is that deforestation rates during the first half of the nineteenth century resulted from the twin engines of economic growth and demographic expansion.

In addition to reforming forestry, Dalhousie instigated one of the nineteenth century's biggest investment schemes by facilitating the expansion of railways. India's overall deforestation rate peaked in the 1850s and '60s because of the massive railway boom started by Dalhousie in the early 1850s. Railway sleepers ate up huge quantities of valuable hardwoods while less valuable timbers fuelled the powerful steam engines as they chugged throughout the hills, valleys and plains of India. Expanding road and telegraph networks contributed to this torrid pace of deforestation.

The rapid pace of deforestation from the 1820s to the early 1850s created anxiety among an eclectic set of botanists, doctors, clergy, military officers and government officials. These individuals began to agitate in the Company's three provincial presidencies: Bengal, Madras and Bombay. They wanted to institute a more comprehensive system of state forest management to slow the pace of deforestation and to ameliorate negative climatic changes caused by the loss of forests. For precedence and evidence, they drew from colonial and European ideas of climatic botany and used local anecdotal evidence of rainfall declines while pointing to the alarming story of the decline of precipitation on the tiny Atlantic island of St Helena that followed after its forests were cut down.

An energetic Scotsman, the physician Hugh Cleghorn (1820–1895) became the unofficial leader of this Indian climate change advocacy group. That a medical doctor led forestry reforms might today seem improbable, but at the time doctors studied botany as a requisite of their medical training. Forestry in India was built from the bottom up and led by medical doctors, often from Scotland, such as Cleghorn. Medical doctors learned botany in their training, and this training led them to take an interest in the plants of India. Though Scotland had a history of tree planting in large aristocratic estates that dated to the eighteenth century, the British Isles did not have a domestic forestry school or a tradition of managing existing forests (as opposed to creating plantations) until the founding of a forestry programme in 1885 at the Royal Engineering College at Cooper's Hill in England. Cleghorn deeply regretted Britain's lack of knowledge in forestry compared to Germany and France, countries that led the world in forestry education and state environmental policy.[24]

Cleghorn first became interested in botany when he started working for the Indian Medical Service in 1841. He became a regular correspondent with Joseph Hooker, who directed Cleghorn to collect Indian flora for Kew Gardens and to inform Hooker's botanical study of South Asia, *Flora Indica*. Like so many Britons in India, the climate and local diseases took their toll on Cleghorn, who applied for and received leave to go to Britain to recover from sickness in 1850. Back in Britain, Cleghorn was welcomed as a leading expert on India's forests because of his activism and connections with Hooker. While Hooker was tramping about India's forests and mountains, Cleghorn attended the 1850 meeting of the British Association for the Advancement of Science, in Edinburgh. There he worked with two former Indian officials and a professor at King's College, London, to report on the climatic and environmental effects of the destruction of forests in India.[25]

The publication of this report in 1851 provided the most up-to-date, comprehensive survey of deforestation in India. It conclusively called for an Indian-wide programme of forestry protection and appealed for the elites of India, who were led by Dalhousie, to protect India's forests, and by implication its climate, from further deterioration. This report encouraged government officials to finally respond to the threat posed by human-caused climate change.

Dalhousie often escaped the steamy, swampy capital of Calcutta. He toured with his military throughout India, including the Punjab campaign in 1849. He also established and frequently travelled to Simla, a hill-station that became the summer capital of high-ranking Indian government officials. Travelling throughout India afforded Dalhousie a personal knowledge of the climate and vegetation of India, and his peripatetic lifestyle aided his conversion to the belief in human-caused climate change. In particular, he was struck by the vast scale of deforestation that he witnessed wherever he travelled, and he realized the importance of trees after seeing the treeless plains of the Punjab, an extremely arid region of India sustained by glacial rivers cascading down the Himalayas.

By the early 1850s, Dalhousie believed in human-induced climate change and began to advocate action to stave off detrimental climatic and environmental changes. In a speech made at the Agri-Horticultural Society in Lahore in 1851, Dalhousie told the audience,

> No power has been more clearly established than this salubrious and fertilizing effect of foliage in an Indian climate. It has been the subject of much enquiry and has been affirmed and demonstrated in every report submitted from different parts of India, many which have passed through my hands.[26]

One of the great questions that Dalhousie wrestled with was how to implement a coherent, functional system of forest conservancy in India that would protect both its forests and its economic growth. The British government in India agreed that the early conservancy of Watson had failed because of the poorly defined boundaries outlining state and private property and its lack of consideration of the rights of landowners and of traditional users. Watson's actions also had no scientific principles because he saw forestry as a job of policing rather than scientific management. But the state, in the view of most scientifically inclined forestry advocates, did need to take a strong stance on defining the boundaries of and regulating state forests. How to do this in one province was difficult enough given the complexities of legal tenure in India. Harmonizing this across the whole of India seemed nearly impossible.

Dalhousie eventually found a perfect chance to instigate legal and administrative forestry reforms when he launched the Company's military into the Second Anglo-Burmese War in 1852. Having already carved the rump of Tenasserim off Burma after the First Anglo-Burmese War in 1824, the Company used the second war to gain control of the teak-rich parts of lower Burma, including the port at Moulmein and Rangoon, famous for its mix of well-heeled, heavy drinking and ambitious European teak-traders. Disorder with the Kingdom of Ava and the king's inability to maintain its treaty obligations led Dalhousie to act. The war also reflected the desire of a small clique of traders in Rangoon and Company officials in London and Calcutta who wanted access to the abundant teak resources in the kingdom. Dalhousie used the military to take possession of the southern tracts of Pegu, and then he forced the king of Ava to give over valuable teak-laden border territories.[27]

Lower Burma became a laboratory, and later the centrepiece, of forestry in India and the entire British Empire. With the acquisition of Burma, the East India Company undertook a survey to better assess Burma's economic potential and potential tax revenue accordingly. Dalhousie commissioned two officers to survey and report on the best method of managing the teak forests of lower Burma. Responding to their recommendations, in 1855 Lord Dalhousie drafted a forest charter that laid the foundations for later developments in India, including the principle that Crown lands not already privately owned were to henceforth be controlled by the government. They were no longer 'wastelands' or 'commons' that anyone would use or abuse as they pleased. This meant that non-private land became public land, to be managed for the public good.

Dalhousie's decision, though justified using a mix of British and Indigenous precedents, fitted with a global pattern of colonial state expansion into environmental affairs during the period. Dalhousie would almost certainly have been familiar with France's decision to declare all forests in Algeria as state property and to establish a forestry service in 1838.[28] He also knew that France, Germany and many other European countries had established forestry departments and schools, some of which had existed since the eighteenth century.

Expansion of Forestry in India

Dalhousie influenced the future trajectory of Indian forestry when he hired a German botanist, Dietrich Brandis (1824–1907), to design a system for managing the wealthy teak forests in Pegu, Burma. British India lacked the expertise required to manage the growth of Indian government forests, and Britain itself, which had been steadily deforested since the Roman era, could not help Dalhousie's endeavours to establish a system of state forestry regulation. Britain had produced a number of world-leading botanists and horticulturalists, usually of Scottish extraction, but prior to the mid-1880s Britain had no schools of forestry.

Brandis hailed from Bonn, Prussia, the European country with the most distinguished reputation for forestry. Although he was a botanist, Brandis had knowledge of German forestry methods, which he modified for the forests in Burma. Brandis's work proved so successful that his system for managing teak continued to be used in parts of Burma into the late twentieth century.

Dalhousie was lucky in selecting Brandis. First, Brandis was not a forester by training.[29] This meant that he lacked the rigidity of Prussian or French foresters who modelled their ethos along military lines. Second, Brandis recognized the need to accommodate

Indigenous forest-dwellers. He listened to villagers and hill tribes, defending them against other foresters in the 1860s and '70s who sought to curtail their access to forest products. Third, he stood up against British and Indian capitalists who sought to plunder forests. Courageously, he battled against the powerful timber lobbies in Rangoon and Bombay, composed largely of profit-hungry British corporations and local wealthy Indians who sought to liquidate India's natural forests for short-term profit.

Forestry reforms almost vanished after Dalhousie left, when his successor, Lord Canning, became the viceroy. Canning overturned Dalhousie's Burmese policy after a rebellion in Rangoon in 1857 and gave over large swathes of the forest to British timber merchants who sought large profits. These British timber barons had the ear of high officials in India and London. Furious over this breach, Brandis wrote to the secretary of state for India in London with an appeal to overturn Canning's decision. It was a bold, perhaps insubordinate, move for a lowly forestry official who was not even British. Surprisingly, it worked. Sir Charles Wood overturned Canning and reinstated the old system of state forestry in Burma, saving its forests from British capitalists, who subsequently looked elsewhere, to independent upper Burma (conquered in 1885) and Siam, to harvest rich stores of teak before the industry could be regulated by the British.[30]

It took decades to make a coherent Indian-wide system of forestry. Initially, Brandis received an appointment of 'special duty' in 1862 to build a forestry department. In 1864 the government established the Imperial Forestry Service, later renamed the Indian Forest Service (IFS), where Brandis held the position of Inspector General of Forests of India from 1864 until his retirement in 1883. In just a few decades, the Imperial Forest Service transformed itself from a ragtag group of people – described in 1871 as a team who, 'beyond a turn for sport and camp life, had no peculiar qualifications for the duties of forestry' – to a service that was recognized as one of the foremost national forestry systems in the world.[31] Brandis guided through important legislation, including the 1865 India Forest Act, the 1878 Indian Forest Act and the 1883 Madras Forest Act.

These acts gave the government the power to proclaim and set aside forests into different categories, including reserved, unreserved and community forests. Foresters could limit access to forest resources, but most local people retained some access, although they frequently protested or broke the law. Today, the question of whether to create strict protected areas, with fences and policing, is hotly contested, but in the nineteenth century these ideas were considered standard by European foresters. Local people fought against the imposition of legislation, as did many other colonial officials who sympathized with them, but foresters took over large, forested tracts of land, which they ultimately controlled. During the imperial period these rules protected forests, but the rules created a rift between local people and foresters. The conflict led to protests, including the deliberate burning of state forests. The forestry policies disrupted both economic and socio-ecological systems, and transformed forests into simply being pieces of a global commodity chain. These developments benefited urban elites who used trade to build personal wealth, but it depressed alternative local economic models that relied on forests. The history of forestry in India parallels that of most former colonies, including settler societies such as in the United States and Australia.

Climate formed an essential pillar of forestry in India, but climatic protection, though important, had a secondary role to timber production. The goal of the Indian Forest Service was twofold: first, foresters sought to maintain a sustainable supply of timber for India's and the global economy and for the use of local people. Whether foresters were effective in creating an equitable and sustainable regime is contested, but foresters thought it was possible to achieve this balance. Second, the IFS sought to protect forests to defend the 'household of nature', which included climate, soil erosion and even animal life. In the minds of foresters such as Brandis, forests could be both harvested *and* conserved.

From India to the World

The individual events that led to forestry in colonial India form part of a larger global change in the second half of the nineteenth century. India was a key site for forestry reform and debates about forest and climate, but it was not an isolated case. India's place in the British Empire and its diverse forest and climate types (from wet tropical to alpine) made it an important example for English-speaking countries and countries with diverse climates. Yet the question of how forests influenced climate was debated and discussed vigorously in almost every European colony during the mid-nineteenth century.[32]

Along with India, forestry reforms in French North Africa, especially Algeria, shaped Indian and global attitudes towards the forest–climate debate in arid regions during the mid-Victorian era. From the 1830s onwards, French foresters argued passionately about the climatic, economic and even moral benefits of tree planting in arid and tropical regions. France had a close-knit group of foresters trained at its national forestry school at Nancy, established in 1824, who worked in France and throughout the French Empire. In the 1860s, foresters in Algeria started to promote the dream of increasing North Africa's rainfall by planting trees. Grandiose claims called for 'resurrecting the granaries of Rome'.[33] Unfortunately, climatic realities and the difficulty of establishing successful plantations of exotic trees meant that this dream, like so many other schemes globally, failed to produce economically viable timber plantations.[34]

From 1869 to 1885, the British Indian government sent its forest trainees to complete their full forestry degree at Nancy.[35] Nancy was not the only forestry school in Europe – in fact, the most renowned were located in German-speaking countries – but more British trainees knew, or could learn, French than German, a language that was not only more grammatically complex than French but less frequently studied in British schools at the time.[36] British officials never felt entirely comfortable outsourcing its education to a foreign country, but the relationship between British foresters and Nancy continued for some time, even after the British established their own forestry school at Cooper's Hill. The reason that the school sent students to Europe to study practical forestry was because Britain had so few examples of large, managed forests. Many of the most influential English-speaking foresters, such as the British colonial forester David Ernest Hutchins (1850–1920) and the American Gifford Pinchot (1865–1946), studied at Nancy.

French and German ideas influenced the IFS, but the IFS retained a strong British Indian culture tied to the larger Anglo civil service in the subcontinent. Indian foresters came from elite British schools, and they had to deal with a complexity of issues that European foresters rarely faced, such as managing forests that were as big as some European countries, dealing with a large country that was host to many different languages and cultures and experiencing severe social isolation while on patrol in remote forests. These experiences meant that other colonial governments sought out Indian foresters for their experience.

Indian foresters – and the precedent of Indian forestry – spread throughout the empire. British Indian foresters were highly prized for their ability to implement new forestry systems in colonial environments. In 1922 Sir Wilhelm Schlich (1840–1925), the second former inspector general of forests, wrote that 'officers have been lent to Ceylon, the Federated Malay States, Mauritius, New Zealand, Australia, South-, West- and East-Africa, the Sudan, Cyprus, the West Indies and British Honduras.'[37] Gregory Barton argues that the spread of Indian foresters throughout the world kick-started global environmentalism: 'Fifty separate forest services protected not only trees but soil, water and – so foresters believed – the climate.'[38] British Indian and French North African precedents influenced the development of forestry in most countries globally from the 1850s to the 1920s.

Forestry departments sprung up around the world in the 1880s and '90s as a result of the adoption of forestry by leading colonies. A few of the notable governments that established forestry departments included those of the Cape Colony and Madagascar in 1881, South Australia and New South Wales in 1882, British Malaya in 1888, the USA in 1891, Thailand in 1896, Japan in 1897 and the Philippines in 1904.[39] The idea that forests played a role in climate amelioration was used by forestry enthusiasts in countries that lacked centralized forestry systems to justify the creation of sweeping forestry laws and new administrations staffed by scientists to manage national forests. Fears of climate change built on top of other anxieties, such as concerns over rapid deforestation and timber shortages.[40]

The 1880s to the early 1890s represented the pinnacle for the belief that forests influence climate.[41] Early evidence from India stoked the belief that forests indeed positively influence climate. Weather stations established by the British Meteorological Office, and earlier by other government officials, in different parts of India suggested that forests that had been protected from fire had an overall increase in rainfall. Commenting to an audience in Germany in the late 1880s, Brandis noted with confidence, 'in the last few years facts have become known which certainly, so far, point to the conclusion that the conservation of forest in several localities has resulted in the increase of mean annual rainfall.'[42]

Foresters proposed that the protection of forests had a variety of positive impacts. In 1900 Berthold Ribbentrop (1843–1915), the third inspector general of forests in India, wrote a history of forestry in India extolling the virtues of forestry and the evils of deforestation. Moreover, his book justified the government's takeover of forests. Ribbentrop argued that deforestation had 'the most deteriorating effect on the climate of India . . . the numerous deserted villages and mounds [indicate] the previous existence of a dense population in parts of the country where cultivation is at present found only in the most favourable stations.'[43] He also justified forestry for its local effects. He wrote, 'There is no divergence of opinion regarding the more local effects of forest in protecting the soil and regulating both surface and subsoil drainage.'[44]

Foresters in India believed they had succeeded in saving India's forests and thus improving its climate, yet aspects of this victory would be short lived. As the next chapter discusses, meteorologists, especially American meteorologists, began to vigorously challenge the argument that forests influence precipitation. Climatic botany helped initiate a global movement to preserve forests in the second half of the nineteenth century, but it played only a limited role in justifying forestry for most of the twentieth century.

4

THE EVAPORATION OF THE FOREST–CLIMATE QUESTION

With a few controversial strokes of his pen, President Theodore Roosevelt established a vast national system of 150 federal forest reserves spread out over 56.5 million hectares (140 million ac). Roosevelt made this bold executive decision partly because he believed that forests were necessary for protecting the nation's rainfall. In his 1908 State of the Union Address to Congress, Roosevelt warned the nation that deforestation, if unchecked, would cause devastating climate change in America like it had in China. Dramatic photographs taken in northern China led him to believe that the country had become a barren wasteland as a result of deforestation. Evoking these images, he warned the nation that 'instead of the regular and plentiful rains which existed in these regions of China when the forests were still in evidence, the unfortunate inhabitants of the deforested lands now see their crops wither for lack of rainfall.'[1] Roosevelt concluded strongly that the state must protect forests in the national interest against those who wanted them privatized or cut down.

Roosevelt's efforts to stop climate change created a national political controversy. The idea that forests influenced rainfall and streamflow divided American politicians, technical experts and the public. Disagreements continued to boil after Roosevelt left office in 1909. In 1910 the U.S. Congress officially investigated the question to inform national policies relating to the regulation of federal forests and waterways. Robert DeCourcy Ward (1867–1931), Harvard's first professor of climatology, complained in 1913 about how climate change had become 'more or less a matter of semi-political controversy' in the country.[2]

Roosevelt dictated the terms of management for 93 million hectares (230 million ac) of public land using executive privilege that allowed him to bypass a Congress filled with statesmen from the western states who would not willingly give away land located within their state boundaries. He used powers from the 1891 Forest Reserve Act and the 1897 Organic Act to put federal lands under the control of the U.S. Forest Service. Gifford Pinchot, Roosevelt's right-hand man and the strong-willed head of the U.S. Forest Service, stoked fear among state leaders that the pair wanted to strip the states of 'their' federal land. Satirists depicted Pinchot as an arrogant monarch making ranchers grovel and prostrate themselves in front of him for a small 'square pole' (approximately one-seventh of an acre), an insignificantly small size for grazers whose ranches could be hundreds of thousands of hectares in size.

Roosevelt's high-handed actions generated alarm among a diverse mix of politicians, farmers, engineers and meteorologists who disagreed with his beliefs on climate. No community expressed more alarm than the constituents of rural western states, who bristled at the loss of millions of hectares of federal land that had been or could be used for grazing, farming and mineral extraction. Roosevelt 'locked' up this land. Critics jumped on any evidence which challenged the forest–climate link to undermine conservation and open resources to the private sector. In 1910 a lawyer in Montana praised a congressional report by the chief of the U.S. Weather Bureau, Willis Moore (1856–1927), which challenged

Roosevelt's views. The lawyer hoped that Moore's report would 'enlighten the public intelligence to the end that so much of the forest reserves as may be sufficiently cultivated shall be open'.[3] If these critics had their way, federally controlled forests would be open to more intense harvesting, farming and grazing.

Discontent over Roosevelt's forest grab finally boiled over into outright political rebellion. Senator Charles Fulton from Oregon led a congressional push to stop Roosevelt from redesignating federal land by proposing an amendment to the Agricultural Appropriation Bill of 1907. The amendment would take away the president's ability to proclaim national forests, leaving that power solely with Congress. In response, Roosevelt commissioned Pinchot and his team to identify potential federal forests in less than a week before the passage of the bill. Roosevelt set aside 6.5 million hectares (16 million ac) in six states a mere two days before the amended act passed.[4] It proved a Pyrrhic victory for a fuming Fulton.

After his conservation coup, Roosevelt's agenda continued with the proposed Weeks Act, which sought to purchase privately owned land to create national forests in the country's east. Unlike America's west, the eastern states had little federal land. To create national forests required the federal government to purchase privately owned land.[5] Advocates of the bill pointed to devastating floods, such as the Ohio river flood of 1907, to justify the creation of protective forests at the headlands of waterways. The belief that forests mitigated flooding was frequently linked with the belief that forests created rain, although some leading foresters, Pinchot being the most prominent, had downplayed the climatic link. The idea retained popular support, not least among Roosevelt.

Roosevelt's progressive conservation agenda once more ran into problems in Congress. Advocates worried that the House Committee on the Judiciary might strike down the Weeks Act by advising that it was unconstitutional for the government to purchase private land in states. A hastily written amendment prepared on the floor added the words 'for the purposes of preserving the navigability of streams' to ensure it would be permitted by Congress under the expansive commerce clause. The argument that forest cover created more regular streamflow underpinned the clause, although Roosevelt's views on the subject suggests he sought to increase precipitation as well. The reworded law passed the Judicial committee.

After the act passed, it was then the engineers' turn to step in and attempt to block Roosevelt. Engineers working for the Army Corps of Engineers argued that forests did not influence flooding or rainfall. With the Corps, Roosevelt faced something of a wounded tiger. The Corps was facing a potential existential crisis because of Roosevelt's multi-use conservation efforts.[6] His Inland Waterways Commission, for example, if fully implemented as he and Pinchot wanted, would take away some of the Corps' authority over the management of national waterways. The commanding officer in the Corps commissioned a sixteen-page report by Major Henry Newcomer (1861–1952) on the relationship between streamflow and rainfall. Newcomer's poorly written report did little to rebut the conservationists' politically dominant position.

Lieutenant colonel Hiram Martin Chittenden (1858–1917), an engineer who worked for the Army Corps of Engineers from 1884 to 1910, produced a more authoritative report that sought to undermine the Weeks Act. Chittenden served for the Corps in Yellowstone National Park twice (1891–3 and 1899–1904) before moving to Seattle, where he first worked as a district engineer before being discharged from the Corps because he could not pass a horse-riding test introduced by President Roosevelt due to ailments.[7] Being forced to retire at the age of 49 because of Roosevelt's whimsical decision embittered him, and perhaps influenced his response. Like many, Chittenden's views on the issue of forests and water had evolved in the 1890s and 1900s. In 1895 he wrote a book on Yellowstone National Park that agreed with the view that forests encouraged water conservation, although he did not speak directly to the question of whether forests increased precipitation.[8] Only two years later, he made known his private dissatisfaction with the same idea to his friend, George Anderson, acting superintendent of Yellowstone. In 1908 Chittenden published an article in *Transactions of the American Society of Civil Engineers* which attacked the ideas that flooding and rainfall were impacted by forests.[9] He argued against the 'commonly accepted opinion . . . that forests have a beneficial influence on stream flow . . . by storing waters from rain and melting snow . . . by retarding the snow-melt . . . by increasing precipitation . . . [and] by preventing erosion'. He challenged the idea that deforestation led to rainfall declines by pointing to devastating floods caused by high rainfall in the deforested eastern parts of the country. He wrote: 'Considering how large a percentage of our forest has already disappeared, the extraordinary rains in all parts of the U.S. during the past year are not exactly in line with this dismal prophecy.'[10]

Never one to back down from a public fight, Roosevelt fought back against Chittenden by using his outgoing presidential address to Congress to criticize the *Transactions* paper. Roosevelt drew on reports from Frank Meyer (1875–1918), a U.S. Department of Agriculture economic botanist who had travelled throughout China for three years exploring that country's flora and collecting useful plants to bring back to the U.S. At a personal meeting, Roosevelt encouraged Meyer to produce an illustrated description of his travels.

The visual images Meyer published showed dramatic images of erosion in Shanxi province. Roosevelt discussed these haunting images in his speech, in which he drew a strong connection between forests and rainfall:

> The climate [of northern China] has changed and is still changing . . . The great masses of arboreal vegetation on the mountains formerly absorbed the heat of the sun and sent up currents of cool air which brought the moisture-laden clouds lower and forced them to precipitate in rain a part of their burden of water. Now that there is no vegetation, the barren mountains, scorched by the sun, send up currents of heated air which drive away instead of attracting the rain clouds and cause their moisture to be disseminated.[11]

Even with this push, the Weeks Act did not end up getting through Congress. Discussion of the bill continued after Roosevelt left office, but the failure to pass it during his presidency undermined the credibility of the view that forests influenced precipitation.

Scientific Challenges to the Forest-Influence Question

Despite the reshaping of global forest policy, no single view on forests and climate commanded a scientific consensus during the second half of the nineteenth century. Even supporters admitted that the idea that forests increased precipitation needed more evidence. In 1864 the eminent American statesman and conservationist George Perkins Marsh (1801–1882) admitted in his study *Man and Nature*, 'Unfortunately, the evidence is conflicting in tendency, and sometimes equivocal in interpretation.'[12] Ambiguity allowed pro-forestry supporters to make a convincing ethical claim that forests should be protected until the scientific debate could be settled. Without disproving the idea, Marsh counselled Americans about the 'necessity of caution in all operations which, on a large scale, interfere with the spontaneous arrangements of the organic or the inorganic world'.[13]

Experts who sought to answer the question faced a litany of challenges. Key aspects of atmospheric dynamics could not even be measured, let alone theorized. Almost all nineteenth-century atmospheric measurements came from near the ground rather than in the upper air, where scientists could study how the higher altitudes shaped wind, temperature and precipitation on the ground.[14] The idea that forests positively influenced precipitation had never been subjected to rigorous experimentation. Experiments only began to be designed in the late 1860s. Reconstructing climate, another possible avenue, ran into dead ends. Analyses of past climate relied on questionable climatic records that for most of the world only began in the 1850s, if not later. A bewildering array of statistical data had to be arranged and new methods had to be devised for determining whether long-term climate trends reflected 'naturally' occurring cycles or indicated actual deviations from the statistical average.

The steady accumulation of data in the late nineteenth century, combined with new theoretical advances at the dawn of the twentieth century, made it possible for experts in meteorology and climatology to begin more firmly questioning the importance of forests in climate and the water cycle. Yet, even with these advances, leading meteorologists and climatologists had a difficult time saying conclusively whether or not forests influenced rain and climate. This ambiguity allowed the question to linger in the United States until the early 1910s.

The intellectual inclination of individuals tended to reflect their professional affiliation (forestry, climatology, meteorology and so on), individual personality and the specific contexts in which they worked. Foresters received significant powers from federal governments by arguing that forests needed protection to conserve climate as well as resources such as wood and water. Giving up the idea that forests had an influence regionally would lessen foresters' authority over the conservation of large forests. Similarly, the Army Corps of Engineers benefited from the view that forests did not influence streamflow. This allowed them to argue more narrowly that riverways were primarily routes of transportation rather than tools for conservation.

This is not to say that all foresters or engineers cynically used arguments, but it does provide political, social and economic explanations as to why foresters supported theories about climate change and flooding risks more than engineers, who frequently argued against this position. The historian Gordon Dodds argues that 'Pressed by their critics . . . forestry advocates, some of whom were privately aware of their own methodological weaknesses, fell back upon enthusiasm and, on occasion, duplicity.'[15] One can apply the same critique to engineers who used whatever evidence they could marshal to argue steadfastly against ideas supported by foresters.

In the end, meteorologists and climatologists were the professionals who principally settled the debate on forests because they became the main authorities on weather and climate. Advances in both fields emphasized the role of continental and oceanic atmospheric dynamics as drivers of regional climate. The emphasis on larger-scale atmospheric processes discredited explanations of weather and climate based on older geographic ideas that gave power to vegetation.

Leaders in both fields became contemptuous of the lingering belief that forests influenced rainfall.

One is tempted to view climatology and meteorology as being more politically neutral because climate experts did not seek to control waterways or forests like engineers or foresters. But both groups had considerable professional interests that impacted on their actions, such as rejecting views that did not conform to their calculations and models. The period from the late nineteenth to the early twentieth century marked a shift in both fields away from 'pluralist science', which included citizen and other expert contributions, towards a more unified, expert-led, monistic approach.[16] This split continues to have lingering ramifications on climate research.

The Rise of Meteorology and Climatology

The rise of meteorology and climatology as scientific thinking undermined key claims associated with climatic botany, although the question of whether forests influence climate lingered in public thought and on the fringes of science for many decades. The historian of science and technology James Rodger Fleming argues that the emergence of these fields encouraged a 'shift from literary to empirical studies of climate, from impressionistic evidence to statements of fact, from dim apprehensions to a recognizably modern climatology'.[17] In the United States, this new science 'put to rest uninformed speculation about temperature changes caused by settlement of the continent'.[18] A similar shifting of the climate guard occurred in India, Germany and elsewhere in the 1900s.

Meteorology and climatology both emerged as distinct scientific disciplines in the late nineteenth century.[19] The Austrian climatologist Julius von Hann (1839–1921) noted in his field-defining 1884 book *Handbuch der Klimatologie* (translated into English as *Handbook of Climatology* in 1898 and 1903), '*Climate* is the sum total of the *weather* as usually experienced during a longer or shorter period of time at any given season.'[20] Hann noted that the goal of climatology was 'to make us familiar with the average conditions . . . as well as to inform us concerning any departures from these conditions'.[21] To do this, climatologists poured over statistics on rainfall, temperature and humidity to determine the average sum of climate throughout the world. Climatological knowledge became an essential part of agricultural science, especially efforts to improve yields by introducing new crops or using new techniques. By the early twentieth century, many U.S. states employed a climatologist for this very reason.

Meteorology had two distinct, overlapping branches. The field is most popularly known as a tool for forecasting weather. Across the world, television meteorologists make nightly predictions about regional and local weather for five-day spans. Weather forecasters draw insights from models and data developed by more specialized meteorological scientists who study atmospheric dynamics in detail. During the early development of meteorology, from the 1850s to the 1890s, most meteorologists had some experience with weather forecasting for the simple reason that governments were the primary employer of meteorologists and they needed short-term forecasts which were necessary, among others things, to plan for extreme weather events, such as hurricanes, tornadoes or blizzards. Longer projections had less pressing importance. It took considerable effort to convince the U.S. Army and other institutions to fund serious theoretical research into meteorological dynamics. After the First World War, meteorology in the United States and Europe fragmented into more applied (forecasting) and theoretical (meteorological science) branches. The new theoretical meteorology was informed by the Bergen school, founded by the Norwegian meteorologist Vilhelm Bjerknes (1862–1951), as well as ideas proposed in the 1900s by the American Cleveland Abbe (1838–1916).[22]

Prior to the mid-1800s, both meteorology and climatology languished compared to established physical sciences, such as physics and chemistry. A mix of amateurs, military officials and experts in a variety of fields all contributed their knowledge to the subjects. This ragtag science left much to be desired in the eyes of those who sought to make them into proper professions. In 1832 the Scottish physicist Jacob David Forbes (1809–1868) lamented in his report on the state of meteorology at the Second Meeting of the British Association that 'Meteorological instruments have been for the most part treated like toys, and much time and labour have been lost in making and recording observations utterly useless for any scientific purpose.'[23]

Meteorology and climatology started to develop stronger professional identities in the mid-1800s during a rapid phase of technological change, growing global trade and expanding European empires. By the 1850s, stations devoted to taking data for meteorology and climatology started to open with increasing frequency in Europe, the United States and Europe's imperial territories. The spread of stations occurred at the same time as the expansion of the telegraph, a technology that had opened up the exciting possibility of relaying weather data quickly ahead of storm systems, giving people time to prepare for incoming weather changes. Weather information could signal an impending nor'easter off the coast of New England, a typhoon about to hit Hong Kong or a tornado-producing storm bearing down on Kansas.

The first meteorologists worked for governments. Vice Admiral Robert FitzRoy (1805–1865), captain of HMS *Beagle* from 1828 to

1836 and founder of the British Meteorological Office, offers a striking example of the opportunities and challenges meteorologists faced at the time.[24] FitzRoy first became interested in forecasting after he nearly sank the *Beagle* during a storm.[25] In the 1850s, he grew increasingly worried by the thousands of lives lost at sea in shipwrecks caused by storms. In 1854 he took up a position at the Board of Trade directing the early Meteorological Office. In this role, he pioneered the daily weather forecast. His daily weather table used up-to-date information from coastal towns about possible storms. For doing this public service, he received praise from seafarers and extensive criticism from the press.

Early forecasters like FitzRoy relied on their experience and observation of previous weather events to predict future events. The proliferation of statistics gathered by weather stations at first overwhelmed meteorologists. A review of FitzRoy's influential *Weather Book* in the British magazine *The Spectator* praised the book but concluded nonetheless that, 'Meteorology has not yet won a place in the ranks of the exact sciences.'[26] Meteorology lacked a firm theoretical understanding of the physical dynamics of weather. Two years after publishing his book, FitzRoy sadly took his own life while suffering from depression, an ailment linked partly to criticism he had received in the press. The Met Office stopped issuing forecasts for over a decade after his death, in part due to 'general incredulity as to the utility of meteorological statistics' and because FitzRoy's reputation had been 'temporarily clouded' by his suicide.[27]

Climatology grew up alongside meteorology as a related discipline but should be thought of as the younger and somewhat weaker sibling. Hann's *Handbook of Climatology* marked the professional origin of climatology. Climatologists did not make claims to be able to predict short-term weather, so they often received less government support. Influential climatological advances came from meteorologists who worked for governments (such as Hann in Austria), but many worked for universities (for example, Alexander Woeikof, DeCourcy Ward and Ellsworth Huntington).

Many climatologists maintained more diplomatic and nuanced views of the question of forests and rain. Hann's personal views on forests and climate shifted from the late 1860s to the early 1900s. In 1869 he said that the belief was 'theoretically well founded' but it had not yet been confirmed.[28] His *Handbook* suggested that tropical forests had a greater influence over rainfall than forests did in temperate regions. Hann's views agreed with those of Woeikof (1842–1916), a Russian polymath who advanced knowledge of climatology, meteorology and geography. Climatologists continued to research forest–climate influences well into the twentieth century, but most came to agree that forests had relatively little climatic influence beyond a small locality.

Meteorologists tended to take a stronger stand against forest–climate questions. This reflected their detailed use of statistics, and after the 1900s, physics and mathematical equations. American meteorologists took a particularly firm line.[29] No one proved a more dogged critic than Cleveland Abbe, the head of the U.S. Weather Bureau and a man described as the 'dean' of American meteorology. In a popular article in 1888 he concluded: 'rational climatology gives no basis for the much-talked-of influence upon the climate of a country produced by the growth or description or forests.'[30] He grew less diplomatic over time. Later he wrote irritably: 'In this day and generation, the idea that forests either increase or diminish the quantity of rain that falls from the clouds is not worthy to be entertained by rational, intelligent men.'[31] As head of the Weather Bureau as well as editor of the *Monthly Weather Review*, Abbe was able to shape the beliefs of other American meteorologists. Henry Allen Hazen, who worked at the U.S. Weather Bureau, summarized the bureau's position at the end of the century in the *Monthly Weather Review*: 'Meteorologists are agreed that there has been practically no change in the climate of the world since the earliest mention of such climates.'[32]

Precipitating the Scientific Debate about Rain

Settling the debate required better knowledge of precipitation, especially its most common variety, rain. Meteorologists and foresters lacked sufficiently accurate long-term information about rainfall amounts and distribution to determine whether variations in rainfall were natural and 'averaged out' across time or reflected actual changes to long-term trends. Naturalists had begun measuring rainfall in London and Paris in the late 1600s but there was scant data for almost everywhere else in the world.[33]

It was not until the second half of the nineteenth century that governments created infrastructures for national weather observations. Tens of thousands of stations were created across almost every climatic region, from the cold Arctic to the steamy equator and from the depths of Death Valley to the heights of the Himalayan mountains. Most stations took daily readings of rainfall, air temperature, pressure, wind speed, humidity and sometimes even gases such as ozone and carbon dioxide. Efforts to establish more meteorological stations, and more reliable ones, in as many locations as possible began to bear some statistical fruit from the 1880s to the 1890s that informed meteorological critiques of climatic botany.

Researchers in forestry, meteorology, agriculture and climatology all founded stations to record observations relating to rainfall.

Foresters, in particular, played a leading role by creating experiments using rainfall gauges to measure the variability of rain inside and outside of forests.[34] Scarcely five years after establishing a trial, Ernst Ebermayer (1829–1908), a German forester, published his first results. Measurements taken twice a day for five years showed that open areas received more rain than forests, but that closed forests had 2.6 times less evaporation. Some of Ebermayer's results actually questioned the supposed link between forests and increased rainfall since more rain fell beyond their range than within them. He warned against seeing the results as 'final and exhaustive evidence', but news spread excitedly throughout the European forestry community.[35]

Foresters used his experiment to claim that forests conserved water and had less evaporation. A few foresters took even more liberal interpretations of his work. Berthold Ribbentrop, the inspector general of forests in India, drew on Ebermayer's results to justify the protection of forests in India for climatic reasons.[36] The results also led to a proposal being put forth at the 1876 International Statistical Congress for greater coordination of research on the forest–climate question.[37] Experiments started soon after in the German Empire, Austria, Switzerland, Italy and France.[38] The early results of other experiments validated Ebermayer's research on the moisture conditions inside forests. A review in 1886 by Woeikof pointed out that all these studies had similar results: '(1) the air and Earth temperatures were lower in the forest than in contiguous woodless places; (2) their variations were less; and (3) the relative humidity was greater.'[39]

Scientists in northern Germany and France challenged part of Ebermayer's findings by suggesting that forests received more rain than open areas and conducted experiments to prove this theory.[40] French foresters established four stations (two in a forest and two in open areas) outside Nancy, France, home of the French national forestry service, to answer the question. Results by the 1890s suggested that 20 per cent more rain fell in the forests than in the adjacent open areas. An even more surprising result came from another German example located near Lintzel, in the north of the country. Lintzel was originally covered in shrubby heath, but starting in 1887 foresters eventually planted more than 3,200 hectares (8,000 ac) of trees. Within less than a decade, Lintzel had received more rain than the surrounding regions. Data from thirteen rain gauges within 80 kilometres (50 mi.) of the site indicated that Lintzel went from receiving less rain than surrounding areas to receiving more than them. The results of most European studies indicated that forests received more rain than open areas.[41]

These European studies were further bolstered by examples from tropical India and Indonesia. In central India, fires destroyed almost 360,000 hectares (890,000 ac) of forest in the early to mid-1870s. After the fires, Indian forestry officials protected the forest from harvesting and allowed forest to regenerate naturally. An 1886 analysis of 22 stations in central India by Henry Blanford (1834–1893) suggested that the rainfall was 20 per cent greater in the ten years after the fire than in the ten years before.[42] Blanford's findings came with considerable weight because he was the official Indian government meteorologist at the time. His interpretation was cited by the U.S. Army's chief signal officer Adolphus Greely (1844–1935) in *American Weather* (1888) and by Hann in the English translation of his *Handbook*. Further experiments for measuring rainfall established in the early 1880s at the Indian forestry school at Dehra Dun showed 'slightly but appreciably higher rainfall in the forest than without'.[43]

In Java, Indonesia, an even more striking example presented itself. The northern part of the island had been deforested by the Dutch for plantations, but the southern part still contained thick forest. Based on prevailing monsoonal winds, the northern part of the island should have received copious rain and the south far less. Yet the south received double the rainfall of the north. The world's authorities on climatology, Woeikof and Hann, both surmised that the southern region's forests were the reason for this increase of rain. Hann wrote, 'it may be concluded, with a good deal of certainty, that, so far at least as the tropics are concerned, forests may actually increase the amount of rainfall.'[44]

Theoretical findings from oceanography seemingly supported this idea. In 1877 the Canadian-born Scottish oceanographer John Murray (1841–1914) proposed in the *Scottish Geographical Journal* that most of a continent's precipitation was recycled back into the atmosphere, where it once again fell as rain.[45] He drew on data from hydrological studies throughout the known world. Calculating the total volume of water in the Earth's rivers using mathematical calculations, he concluded that only one-fifth of all precipitation returned to the sea. A prominent German glaciologist, Eduard Brückner (1862–1927), refined Murray's theory and reinforced his findings. Brückner is widely recognized today among climate change researchers because he attempted to reconstruct the historical climate to understand glaciers and ice ages.[46] Brückner wrote: 'A water particle, which came to the country from the ocean through the atmosphere, falls an average of three times as precipitation, before it returns to the bosom of the ocean.'[47]

Findings from Europe and the tropics, though provocative, came under scrutiny from meteorologists around the world because of their methodological flaws. Even today, designing a proper experiment is so difficult that in many contemporary scientific fields a surprisingly large proportion of studies use improper methods or cannot be replicated.[48] Think then how much trouble researchers in the late nineteenth century faced when trying to devise their experiments. Agricultural and forestry research, despite centuries of innovation, still used relatively simple experimental methods until well into the twentieth century. It was not until the publication of Ronald Fisher's (1890–1962) pioneering *The Design of Experiments* in 1935 that foresters and agricultural researchers had a more solid statistical foundation on which to design research experiments.[49] Every early experiment to measure forest influences would have failed according to Fisher's methodology.

Most forest–climate studies did not meet the methodological standards of the time for meteorology or climatology. Technology and statistics for measuring rainfall improved during the mid-nineteenth and early twentieth centuries but intractable problems remained.[50] The rain catch from each gauge offered a snapshot of rainfall in one location, but technical limitations and the variable distribution of rainfall across a landscape meant that it did not necessarily reflect total rainfall over an area or even a location. Weather researchers first had to identify problems before they could try to compensate for errors.

No country advanced knowledge of variations in rainfall distributions and the challenges inherent in collecting rain more than Britain. In many respects, Britain offered an ideal lab for learning about rain. Most of the British Isles received consistent rainfall and the islands had diverse meteorological conditions, especially strong winds, which helped to test the flaws in gauges. It had the world's best postal and rail systems, so information could be cheaply relayed. The country also had a rich tradition of amateur meteorology dating back to the 1600s.[51] Amateurs continued to play an important role in climate research until the early twentieth century.[52]

George James Symons (1838–1900) used all the advantages that Britain had to offer to build up a vast private network of rainfall observers, beginning in 1859 while he was only 21 years old. Symons's father died early in his life, but he scraped enough money together to study at the Royal School of Mines after leaving school. His ability and interest in weather drew the attention of Robert FitzRoy, who hired Symons as an assistant in the UK Meteorological Office. FitzRoy eventually had to let Symons go from the department because he was only interested in working on rain, an important albeit single dimension of weather, and thus his self-enforced specialization stopped him from fulfilling his wider job. This passion for rain characterized Symons's life. After leaving the Met Office, Symons created a vast British Rainfall Organisation using Britain's postal system that, at its peak, included approximately 3,500 rainfall-tracking stations.[53] Symons became a world authority on rain gauges for a variety of reasons, not least because he corresponded with so many people who tested out different rain-measuring systems. He was the person who invented the standard rain gauge used by the British Met Office.

Symons encouraged the development of better meteorological measurements in the British colonies and in Europe.[54] His book *British Rainfall 1868* laid out rules for collecting rainfall that many colonial governments followed. For instance, he suggested that all rainfall measurements in a country be taken at 9 a.m. to standardize data. Symons particularly singled out India for its poor records. Like other British scientists, he believed that India should be a world leader in the study of rainfall because of its diverse climates and large size.[55] Unfortunately, observations before 1875 could not be taken seriously. Symons wrote that in the past: 'Indian rain gauges were taken indoors at night and locked up for safe-keeping.'[56] After 1875, his gauge and the methods for recording laid out in *British Rainfall* took hold in India as well as in other colonies.[57] In an 1876 handbook for observers, Blanford advised that Symons's gauge was 'the most convenient and trustworthy'.[58]

Several key conclusions emerged from the first generation of more standardized rainfall data. First, the increase in stations around the world enabled researchers to conduct more experiments and collate more data from diverse regions; in doing so, they realised conclusively that rain fell unevenly across landscapes. In some places, people might experience a storm with a downpour of rain in one area while a neighbouring area is completely dry. Random disparities in topography, such as a mountain that casts a unique rain shadow, also mean that proximate locations can vary significantly from each other. Only a large network of rainfall gauges could provide enough data required to 'smooth out' the errors inherent in variations.

Second, the scale, in terms of time and space, of late nineteenth-century studies was too small to make any meaningful inferences. Almost every experiment had an insufficient sample size, and therefore innate errors. The Lintzel study relied on fourteen stations, Ebermayer seven and the Nancy experiment only used four. These stations could not have possibly accounted for the natural variability of rainfall across time or in a place. In Nancy, the experiment stopped collecting data. France's leading official meteorologist warned against using the data from Nancy because of its limited observations and timeframe.[59] In Lintzel, a different set of problems was identified. The location of observations changed part way in the study.[60] The growth of trees itself could have explained the increase of rainfall because forests often received more rain merely because the lack of wind meant rain fell straight down rather than at an angle and thus proved more difficult for gauges to catch.

Third, meteorologists began to call into question the usefulness of statistics to explain the principles of atmospheric dynamics, including precipitation. Abbe cautioned about the problems inherent in rainfall observation: 'the irregularities in the distribution of rainfall . . . are far greater even than the irregularities in temperature, so the index of variability becomes correspondingly large.'[61] These errors meant that 'disappointment awaits those who would demonstrate climatic changes therefrom.'[62] By 1910 meteorologists had stopped seeking a statistical solution. Moore, Abbe's successor, wrote to Congress: 'trustworthy records of temperature, of rainfall and of other meteorological elements do not cover a sufficient range of time to furnish all the data necessary for a *statistical* solution.'[63]

Death by Government Commission

Whether forests created rain was a question subjected to a series of government commissions in the 1900s by different countries. Commissions in the United States and India shifted prevailing government forest policy away from rainfall modification and towards water and soil conservation. Foresters began to emphasize how forests conserved soil and water as a key justification for why the government should control large forests. In the following decades, the link between forests and water conservation fell out of scientific favour, a topic discussed in the following chapters.[64] The belief that forests influenced climate did not die in the popular mind, or even in the eyes of many foresters, but it no longer attracted the same level of scientific attention or government support.

Foresters in British India had long touted the climatic benefits of forest protection. These claims relied heavily on European examples as well as reports from central India and Java. In 1906 Dr John

Nisbet (1853–1914), a former forester in India, asked John Morley (1838–1923), the secretary of state for India from 1905 to 1910, to instigate a proper government commission of the question as it related to India.[65] Nisbet argued that the subject had never been seriously investigated by the Indian government despite it being a widely held belief among many foresters and high-ranking British India officials. Morley agreed to investigate the question. The government's bureaucracy creaked into action, instigating an inquiry into the question.

To help determine whether forests truly influenced precipitation, the India Office first solicited the advice of Sir Wilhelm Schlich, the highly respected German forester who had previously directed the IFS. Schlich noted that even the most meticulous research in Europe by German and French foresters had given no definitive answer to the question. Personally, Schlich doubted that forests played a major role on the subcontinent, a view that ran contrary to many of his forestry colleagues in India. The question was so vast he believed that it would be difficult to properly assess. Nonetheless, the government in India went ahead and sent circulars to all the provinces. The document contained questions about rainfall over the previous fifty years. Two such queries were:

> Was there any reason to believe that during the last half century the amount and distribution of the rainfall over large tracts of country had altered permanently for the better or worse?
>
> Where such change had taken place, was there any reason to connect it with the destruction of forest vegetation in the catchment areas of the rivers and streams?[66]

The response from the majority of respondents, including the largest governing administrations in Madras and Bombay, replied that there was no evidence of declining rainfall.[67] The replies highlighted the difference between anecdote and reliable data. Sir Charles Bayley, lieutenant governor of Eastern Bengal and Assam,

> remarked that the question of rainfall was one which he had constantly discussed with native gentlemen, and especially with elderly men, and he had no hesitation in saying that, at any rate, in central India, and Hyderabad . . . there was a general impression that the rainfall had largely decreased within the memory of living men.[68]

Ironically, statistical data on rainfall showed the opposite: rainfall had increased in those areas.

Gilbert Walker, director general of observatories in India, reviewed all existing rainfall and meteorological data. He concluded that evaporation from forests contributed only a paltry 5 per cent of the subcontinent's rainfall. His view disagreed with that of his predecessor, Blanford, who linked increased forest cover with rain. Walker wrote, 'The direct influence of forests upon rainfall is, I believe, almost universally regarded by specialists as small.'[69] He referred to Hann's work, and did even note that, in some tropical regions, forests might have more influence over rainfall, but he concluded from the evidence that rainfall in India showed no correlation with forest cover.

Walker disagreed with Blanford's original views that afforestation in central India had led to increased rain. He felt that, in regards to Blanford's observations, 'it is clear that the periods are too short' to gather conclusive proof.[70] He further pointed out that rainfall in the southern Central Provinces and eastern central India fell after 1885, a view which invalidated Blanford's argument that the regrowth of forests increased rainfall in the area. Overall, the trend for all of India was one of variation that could not be correlated with forest cover. Walker concluded,

> although the information at my disposal regarding the destruction and re-establishment of forest is extremely scanty, it would suggest that destruction on a fairly large scale was in vogue some fifty years ago, and that during the past twenty years a material improvement had taken place. The effect of this would, however, have been a relatively low rainfall forty to fifty years ago and there would have been a gradual improvement during the past fifteen or twenty years – a result entirely different from that observed.[71]

Walker's main research focused on the monsoon, an atmospheric system located in the Indian Ocean basin that India depended on for rain. He argued that variations in the monsoon better explained changes in local rainfall in India. Walker sought to understand why in the 1890s and early 1900s India received five weak or delayed monsoons. This compelled Walker to look to other measurements throughout the Indian Ocean region, including the level of the Nile, atmospheric pressure in the southern Indian Ocean and sunspot cycles, to understand what drove India's seasonal shifts. He concluded that the height of the Nile correlated with rainfall in India during the period 1888 to 1907, an observation that suggested that increased or decreased rainfall in India had little to do with forest cover. In conclusion, Walker noted, 'I believe that the effect of forests on rainfall is small, and probably does not reach 5 per cent.'[72]

Walker's findings fundamentally challenged the longstanding official position in British India that forests increased rainfall. V. K. Saberwal suggests that many Indian foresters continued to adhere to the idea that forests increased rain, but this does not mean that the idea that forests influenced rainfall justified government forestry policy.[73] The Indian government maintained its strict control over forests for a variety of reasons, but climate was a low priority that foresters only invoked now and again, in journals such as the *Indian Forester*, which had little policy impact. The historian James Beattie suggests a wider decline of the forest–climate connection throughout the British Empire in the early 1900s, a dating confirmed by this book.[74] Certainly many individual foresters, in India and elsewhere, continued to advocate a belief in climatic botany well into the twentieth century. Yet these increasingly fringe views usually came from lower ranked foresters and did not influence government policy.

A 1909 inquiry by the House of Representatives Committee on Agriculture in the United States came to a similar conclusion. Charles F. Scott, the chair of the committee, commissioned the opinion of Weather Bureau chief Willis Moore to advise Congress on the question. Moore's report challenged ideas that forests influenced rain, and forests regulated streamflow.[75] Moore's views reflected the established American meteorological position that forests had little influence on climate. By positioning himself as a friend of conservation, Moore tried to sympathetically dismiss the idea. He argued, 'it is doing an injury to a good cause to attempt to bring to its support the false reasoning and mistaken conclusions of enthusiasts.'[76] A *Washington Post* headline summarized his view as: 'False Reasoning Injure a Cause of Conservation'.[77]

The Weeks Act was eventually passed in 1911 and President William Howard Taft (1857–1930) signed it into law. A provision in the revised law required that a National Forest Reservation Committee composed of three congressional secretaries, four congressmen and the Geological Survey had to agree to any land purchase. It also set forth a quantitative assessment that would make it more difficult to approve forest reservation by the federal government. The passing of the Weeks Act, in the words of one historian, 'marked the high point of uncritical use' of arguments about forests, streamflow and climate in the United States.[78] Still, the law achieved its intended purpose: from 1911 to 1985, the Weeks Act was invoked by the government to purchase land, creating 45 national forests in 24 states stretching from Texas to Maine. The Weeks Act, like much environmental legislation, was applied far more widely than the intentions of the initial drafters. Ironically, the positive legacy of the Weeks Act happened despite the law's increasingly contested scientific justification for protecting forests to conserve water.

The view that forests had little impact on regional or global climate was maintained by the world's leading meteorologists and climatologists from the 1910s until the late 1970s. Climatologists pointed to exceptions, such as the wet tropics of Southeast Asia and Brazil, but this exception received little scientific attention until the question of forests and the climate of the tropics rose to prominence towards the end of the century. In some parts of the world climatic botany retained more public credibility but again, these locales did not shape orthodox scientific views. The dominance of theoretical models and the downplaying of human action is one reason why climate researchers expressed more public alarm at the possibility of global cooling and the return of ice ages than global warming prior to the 1970s. There has been a resurgence of interest in anthropogenic impacts on climate since the 1970s, the decade when climatic botany once again became a subject of serious scientific interest.

Not all foresters agreed with this climate orthodoxy. The respected U.S. Forest Service scientist Raphael Zon (1874–1956) summarized his professional views in a 1927 scientific review of research on forests and climate: 'The influence of forests upon climate has been a subject of investigation for a long time, and is not settled yet. Now and then this influence has been exaggerated, thus leading to the other extreme of denying it entirely.'[79] Zon's attempt to defend the U.S. Forest Service against criticisms (most of which were later validated) failed to change scientific attitudes on the topic, but Zon and other writers, such as the colourful forester and popular author Richard St Barbe Baker, discussed in the next chapter, helped to keep these ideas alive into the mid-twentieth century.

5

SAVING THE WORLD FROM DESERTS

'Here is a state of emergency, a state of war. The great Sahara desert is invading Africa along a two thousand mile front at the rate of thirty miles a year in some places.'

Richard St Barbe Baker, *Sahara Challenge* (1954)[1]

The idea that forests regulated climate reached its highest levels of popularity – and alarm – with prognostications that deforestation would lead to the progressive aridification of the atmosphere, vegetation, soil and eventually human life itself. Popular scientists in the late nineteenth century started to raise a new environmental spectre that titillated the public: expanding deserts that gobbled up agriculture and civilizations. The threat and fear of the spread of deserts, or 'desertification', as it was coined in the 1920s, helped to establish international scientific and public 'geographic imaginaries' of climate change.[2]

The expansion of European empires and the rise of speculative science combined to create contradictory conjectures about Earth's future. On the one hand, alarms were raised about desertification and its potential impacts on vulnerable fertile regions; on the other, there was hope that science and technology could restore ancient climates and reclaim what deserts had once taken. Numerous scientists used media, popular science, literature and politics to promote the hope that deserts could be reclaimed into Edenic climates. Advocates pointed to examples from ancient texts, the surface of Mars and contemporary science to marshal a powerful, and at times persuasive, vision of Earth's decline into desert. Many of the people who supported these ideas had a penchant for the grandiose, and they often went against prevailing scientific opinion, but they had a wide influence in shaping colonial and international programmes for managing deserts. Efforts to push back against desertification proceeded in the Sahara, the Kalahari, the American West, Australia and the Middle East.

Desertification must be situated within the context of catastrophic thinking of the era. The mid-nineteenth-century discovery that the Earth had undergone one or more ice ages presented the possibility that much of Earth, especially northern Europe and much of North America, could be once again covered by snow and ice. Other people feared that Earth could end up a barren desert like our planetary cousin Mars. In the twentieth century, along with the return of ice ages, other potential end-time scenarios included Earth being hit by a meteorite and later, during the Cold War, fears of a nuclear winter.[3]

Debates near the turn of the twentieth century about the climate and vegetation on Mars reflected new anxieties about Earth's own climatic future. Research and advocacy by Percival Lowell (1855–1916), an American amateur astronomer and mathematician and founder of the renowned Lowell Observatory in Arizona, fuelled fears that Mars's present was Earth's future, and that the expansion of the deserts, along with concurrent loss of rainfall, would lead to our planet's climatic and civilizational collapse.[4] Lowell came from a prominent Boston family that had made its fortune in cotton manufacturing. Though he proved a brilliant mathematics student at Harvard, he never lived up to family expectations. He became engaged to Rose Lee (1860–1953), the sister of Theodore Roosevelt's first wife Alice Hathaway Lee Roosevelt (1861–1884), but broke it off, earning him the hostility of the northeast elite, not to mention the future president. Seeking solace, he undertook long visits to Japan, where he sampled the intricacies of Japanese culture and art with a select cadre of aesthete expatriates. Interestingly, the years he spent in Japan led him to conceive of an idea of Eastern culture that he then applied to the planet Mars. In his Orientalist book, *The Soul of the East*, he predicted that Eastern cultures had run their natural course, and were dying of old age though they retained exquisitely preserved culture, manners and art. In Japan, as in the case of the Moon, he wrote, 'we behold . . . the spectacle of a world that has died of old age.'[5] These views mixed his Orientalist appreciation of Japan with progressive Western attitudes that saw 'pre-modern' cultures fading into extinction.

Lowell applied the same reasoning to all planets, including Earth. Like human cultures, planets ran their natural course and died out. When he learned that Mars would in 1894 pass close to Earth, he funded a new telescope to be built just outside Flagstaff, Arizona. He commissioned a leading American lens crafter, Alvin Clark & Sons, to cast and grind a lens that was 60 centimetres (24 in.) in diameter. Lowell housed the telescope, one of the largest built at the time, in a large dome built of ponderosa pine overlooking Flagstaff, a small town that emitted very little interfering light, meaning that it was a perfect point from which to observe the night-time sky. Completed in 1896, the new telescope spanned close to 10 metres (33 ft) and weighed 6,000 kilograms (6 tons). Lowell made observations personally and dedicated his life to unravelling the mysteries of the desert planet. He scoured the surface of the planet for long hours over months and years, furiously sketching what little he could see through adverse weather conditions, both on Earth and Mars.

Slowly he sketched maps that revealed geometric lines spanning the planet. The uniform nature of the lines on the surface could mean only one thing to him – these were not natural features but the work of intelligent beings. His findings confirmed the sightings of the Italian astronomer Giovanni Schiaparelli (1835–1910), who had also mapped channels on the surface of the red planet. Lowell saw Earth's future on the surface of Mars. Reasoning that planets slowly lost hydrogen and moisture to outer space, he proposed that all life on all planets faced a slow, deadly process of desiccation. He conjectured that Martians responded to the steady desertification of their planet by building canals that piped water from the melting ice caps to the warmer equatorial regions which hosted their advanced civilization. The construction of these canals dwarfed all human efforts at infrastructure – the Great Wall of China, the pyramids and all modern cities and roads. These canals spanned the entire Martian globe and required planning, resources and technology unfathomable on Earth.[6]

Lowell advanced his views at the highest levels of the profession. In 1910, at the British Astronomical Society, Lowell argued that the new observatory in Flagstaff gave the keenest view yet of the red planet. The 'curious markings' on Mars had been confirmed by more than fifteen years of observations. The canals drained the melting polar ice sheets and fed the water into a series of what from Earth looked like dots, which he believed most likely represented a chain of oases, 11–24 kilometres (7–15 mi.) apart, or possibly a continuous belt of vegetation. Either way, the canals were artificial and clearly in his mind created by intelligent beings.[7] Further, the vast global scale of the canals indicated that Martians were not wracked by destructive nationalism to build massive, planet-spanning infrastructure that guaranteed their survival as a species. These were not historical structures only, or ruins. Over the course of his observations, he had seen that new canals appeared in a region where previously they had not been observed. Clearly the building of canals was ongoing and a race against time by the planet's savvy inhabitants. 'You should know that the lines you will see [in his presentation of photographic plates] were certainties, not matters admitting of the slightest question.' Moreover, the canals were not to be understood as large ditches, but 'artificially fertilized strips of country, connected with, and vivified by . . . the melting of the polar cap'.[8]

Newspapers from around the world picked up on Lowell's evidence for an advanced Martian civilization fighting for survival against the expansion of deserts. Novelists rushed in where scientists feared to tread and fleshed out the imaginative contours of Lowell's discovery for an eager public. The idea of a dying planet inspired generations of science-fiction writers, from H. G. Wells to Ray Bradbury. A contemporary of Lowell's, Robert Cromie, laid out a fictional account of life on Mars that received the stamp of approval from Lowell himself. In the novel *A Plunge into Space* (1890), Cromie painted a description of the surface of Mars that predicted a horrific future for Earth. When his fictional characters arrive on the red planet and exit their spaceship, they behold 'a boundless plain of fine red sand', which is

> stretched round them as far as the horizon on every side, east, west, north and south. In this dead waste there was neither hill nor dale, mountain nor lake, nor bird nor beast. Neither was there any living thing whatsoever, animal or vegetable: not a

> shrub, not a leaf, nor even a blade of humblest herbage upon it, and over it a dull red sky hung gloomily, unstirred by a breath of wind, unrustled by the pinion sweep of birds – a fitting landscape for the well-naked planet Mars, the god of war, desolation and death.[9]

This imaginative projection of a desert planet as a potential future for Earth added momentum to many in the global conservation movement – and to many professional foresters as well. Colonial forestry officials saw the same desert future for every region of Earth if humans did not act to protect its forests and therefore its climate. In 1907, well before the theory of canals on Mars had been discarded, one forester observed in the prominent forestry journal the *Indian Forester* that the Martians would lose all their supposed oases (the dark spots that telescopes made possible to see on Mars) to desert (the light spots). Humans had to learn from their mistakes and initiate forest-conservation policies on Mars like the British were undertaking in India.[10]

Alfred Russel Wallace struck the first major blow against Lowell's theory with his 1907 book *Is Mars Habitable?*[11] The public esteemed Wallace as a scientist, second only to Charles Darwin himself. Wallace reminded readers of the obvious – that Mars is further away from the Sun than the Earth and contained a very thin atmosphere. Since the atmosphere insulates the planet from the cold of outer space, the thin atmosphere on Mars could not possibly warm the planet enough to support life. Lowell's suggestion that the equatorial regions of Mars boasted a climate like the south of England was ridiculous. Wallace conceded, however, that the lines observed on the surface of Mars were nothing short of amazing – they stretched the same distance as London to Bombay. He also agreed with Lowell that 'no natural phenomena within our knowledge showed such regularity on such a scale' in reference to the canals.[12] But because Mars had a different geological history to Earth, the explanation for their origins could still be a natural one, Wallace countered. The theory that Mars was formed by the aggregate of meteorites would imply structural differences with the Earth, and perhaps its crust had stress fractures, as seen in the lines observed by Lowell. Wallace concluded that the Harvard physicist and astronomer W. H. Pickering had been right all along when he argued in 1904 that the canals of Mars were rifts in the volcanic crust, caused by 'internal stresses due to the action of the heated interior'.[13]

After the publication of Wallace's book, the theory promoted by Lowell dissipated with every passing year. New techniques of spectroscopic analysis indicated that Mars contained very little water on its surface. Finally, as newer telescopes did not reveal any lines on the surface of Mars, evidence mounted that the lines observed were merely optical illusions created by imperfect telescopes. However, the theory that intelligent beings created canals on Mars continued to linger in the minds of many – that is, until the unmanned probes Viking 1 and Viking 2 were launched in 1975 on missions to Mars, taking photographs of its surface. The footage captured decisively buried the theory.[14] They revealed no canals whatsoever, but they did reinforce the fact that deserts covered the surface of Mars, and therefore stoked the fear that Earth's future would match that of our closest neighbour in the solar system. The idea that Earth faces the same future as Mars because of our 'wicked' ways remains popular even in the twenty-first century.[15]

Deserts, Deserts Everywhere

Approximately one-third of the surface of the Earth is classified as a desert, a term that refers to an arid place. The Afro-Asian belt of desert constitutes fully half of the deserts on Earth. From the Atlantic shores of Africa in the west and almost to the very waves of the Pacific in China, a vast, single arid zone stretches 19,300 kilometres (12,000 mi.). From space it appears as a single desert that encompasses North Africa, the Middle East, Central Asia and northwestern China. The Sahara in North Africa, the Arabian Desert that stretches from Egypt to Iraq, the Dasht-e-Kavir in Iran, the Indus Valley Desert in Pakistan and India, the Kara Kum in much of Turkmenistan, the Lop Desert in northwestern China and the Gobi Desert in western China all form a chain of sand, pebbles and rocky outposts between the 20th and 45th parallel. In the western United States and Mexico, the deserts include the Great Basin, the Mohave, the Chihuahuan Desert and the Sonoran Desert.

From space you would see in the southern hemisphere a global belt of dryness across Australia, the southern cone of South America and parts of South Africa. The Kalahari, the Atacama, the Patagonian Desert and the great burning deserts stretching across Australia's red interior all cluster around the 20th to the 30th parallel. It is less dramatic, perhaps, than the deserts found in the northern hemisphere, because so much of our southern globe is water with comparatively little land, and yet this fact makes the deserts of the southern globe in many ways more dominant as a geological feature.

Deserts hold an enigmatic place in the imagination of those who live outside of them. Urban peoples have often expressed unease about those who lived in deserts and arid places. Ibn Khaldun (1332–1406), the great fourteenth-century Arab scholar, offered a cyclical theory of history in his book the *Muqaddima* which suggested that, over time, nomadic peoples of arid lands would raid and displace sedentary people living in cities. Khaldun believed that conquering nomads eventually urbanized and led agrarian lives, before being conquered by other desert tribes, beginning the cycle anew.

Khaldun lived in a dynamic North African and Middle Eastern political environment that saw the emergence of new empires by once nomadic peoples such as the Bedouin and Mongols.[16] Those who lived in or near deserts tended to lead nomadic lives because it was easier to rely on domestic animals, natural resources and trade (often associated with weaving) rather than relying on farmed crops for a livelihood. Desert-dwelling tribes could be fierce enemies for urban agrarian people. Even today, desert dwellers are some of the most feared soldiers in the world. Afghan Pashtuns, who are adapted to the country's arid, mountainous terrain, have struck fear into the hearts of British soldiers in both the nineteenth and twenty-first centuries.

A romantic vision of deserts emerged in literature and the arts under the auspices of European Orientalism following France's invasion of Egypt in the late eighteenth century. Western Orientalist interpretations grew in popularity in nineteenth-century France and Britain through stories of European colonial expansion and exploration. These views have been fed, to this day, by a steady media infusion, from the 1962 film *Lawrence of Arabia* to the Star Wars universe. For the latter, creator George Lucas designed the aesthetics for Luke Skywalker's home planet, Tatooine, and his mud home based on the similarly named Tunisian city of Tataouine.[17]

In the eyes of many scientists in the past, the desert was not sequestered into a romantic past or future, and nor was it a purely fictional threat. Starting in the mid-1800s, a number of European foresters, geographers and colonial officials began to propose a frightening scenario for the future of the Earth: a progressive expansion of the world's deserts that would destroy modern civilizations much like it had those in the past. Desertification, in the minds of these Western and colonial scientists, equalled a devastating loss of climatic stability and biological productivity. The fear of desertification lasted into the last quarter of the twentieth century. A 1977 United Nations (UN) conference on desertification described the issue as 'the intensification or extension of desert conditions . . . leading to reduced biological productivity, with consequent reduction in plant biomass, in the land's carrying capacity for livestock, in crop yields and human wellbeing'.[18] Even today, the ghosts of desertification remain embedded in international and intergovernmental environmental programmes for countries located in arid regions.

The Sahara

Looking down at the Sahara from an aeroplane, you see an ocean of sand – an endless horizon of yellow and white that reflects the sun, with eruptions of rocks, cliffs, ridges, salt plains, pebbles and sand dunes. Rarely, you might spot a green oasis. It is easy to think of this desert, and so many other desert areas, such as Death Valley in California, or the Gobi Desert in northwest China, as unchanging eternal features of the landscape. In the Sahara, the vast stretch of sand seems eternal. Nothing could be further from the truth. Now we know that just as the ice age advanced and retreated, so too did the deserts advance and recede, including the greatest sandy desert of them all, the Sahara. During the Last Glacial Maximum, 26,000–20,000 years ago, the Sahara was even larger than it is today. That's remarkable to consider, as you could drop the continental United States right into the centre of the Sahara and neither coast would touch its edges.

The exploration of the Sahara slowly revealed a massive graveyard of cities and ecosystems. Traces remain of once verdant valleys, rivers, lakes and swimming holes. What seems unchangeable has experienced massive shifts in climate, and supported, for the ancient world, sizable populations. These discoveries did much to influence the nineteenth- and twentieth-century idea that climate changes over time.

Deep into the central Sahara, early human species lived and thrived up to 400,000 years BCE. Sometime around 70,000 BCE the Sahara began to dry out, only to rise in humidity again around 10,000 BCE, around the time when hunter-gathers began to populate it as the glacial ice sheets in the north began to retreat. The period between 10,000 and 5,000 BCE marked what archaeologists of the Sahara call the Wild Fauna period, when rock art proliferated. The Cave of Swimmers gives tantalizing clues to the once fertile region of the Sahara. Discovered in southwestern Egypt in the mountain chain Gilf Kebir in 1933, it contains rock paintings of human figures diving and swimming. It is estimated to have been drawn 10,000 years ago. Wild Fauna art reveals a full range of typical African megafauna: elephants, giraffes, buffalos, rhinos, lions and hippopotami. The glaciers left behind lakes that slowly drained into perennial rivers and streams scattered throughout an abundant savanna dotted by pockets of forests, whose petrified remains are clear even today in the southern reaches of Libya. People from the Wild Fauna period etched their artistic work deep into the stone with flint tools, drawing oversized animals alongside small figures of human hunters, indicating perhaps their awe of the powerful animals, which they possibly considered even divine. Later drawings from different cultures influenced by the ancient Egyptian and Mediterranean worlds revealed houses, chariots and deep pools with swimmers diving and cavorting.

The lost cities of the Sahara are briefly mentioned in ancient Greek and Roman sources. It is only recently that it became possible to study the Garamantian empire in southwest Libya once again, since the fall of the Libyan ruler Muammar Gaddafi in 2011. The founders of this lost empire are described by Herodotus as peoples who traded from the centre of the Sahara. They left behind thousands of miniature pyramid tombs, hundreds of miles of tunnels, and underground waterways that tapped into the underground aquifer, as well as cities with populations estimated to be almost 10,000 people – very large for the ancient world. They farmed, hunted and rode chariots from what is today the Fazzan province of southwest Libya, basing their agricultural civilization on a series of depressed valleys that tapped into the aquifer at higher elevations. They traded semi-precious gems north to the Roman cities along with gold from Africa and African slaves sold to the Mediterranean cities on the northern coast of Africa. Two things brought them to an end: the Sahara, advancing north, and the Romans, advancing south to fight brigandage.

As the evidence mounted in the late nineteenth and early twentieth century that civilizations once flourished in the sands of desert regions around the world, alarming questions about climate arose. Would the desert spread northward, into an already increasingly arid Mediterranean? Was the advancing desert human-made? Would sub-Saharan Africa sink under the waves of the expanding dunes?

The lakes that once lay on the surface of the Sahara captured the imagination of the mid-Victorian era. One of these lakes was large enough to constitute an inland sea and it connected to the Mediterranean Sea. The European discovery of these lakes gave rise to grandiose plans for flooding the great desert and turning it once more into a great extension of the Mediterranean. The famous geologist Sir Charles Lyell (1797–1875) was the first to argue that the features of the Earth were not created in their present condition, but moulded over aeons of time, and he believed strongly that the Sahara had once been a sea:

> The great African desert, or Sahara, was submerged in the modern or post-tertiary period . . . the fact that the Sahara was really covered by the sea at no distant period has been confirmed by many new proofs . . . the Sahara was under water . . . [and] extend[ed] over the vast space from West to East in the desert.[19]

Lyell also mentioned inland sea cliffs, and 'old sea-beaches covered with shells. The ancient sea appears once to have stretched from the Gulf of Gabes, in Tunis, to the west coast of Africa north Senegambia, as wide as 800 miles.'[20] Drawing on snippets of ancient sources, enthusiastic foresters speculated that the Sahara was once a well-watered breadbasket for the world.

British and French colonial agencies saw Africa south of the Sahara as particularly vulnerable to desertification. UN agencies and national governments continued this vision. Colonial and later national governments as well as international agencies proposed, and at times tried, many plans to stop the advance of the desert by replanting it with trees.[21]

In the late 1800s, the French and the British concocted elaborate plans for turning back the clock of time on the Sahara. In the 1870s, the French War Ministry charged a civil engineer, Captain M. E. Roudaire (1836–1885), with conducting a topographical survey of the western portions of the Sahara between the Gulf of Gabes and the Tunisia–Algeria border. Roudaire argued that much of the Western Sahara lay below sea level. In this inhospitable interior he found a series of seasonal salt lakes and marshes in a chain of depressions, or *wadi*s, that is now known as Chott El-Mehrir. From his geographical survey and reading of classical sources, Roudaire surmised that these seasonal lakes had once formed a mighty inland sea that slowly dried out over time. He published his findings in the *Revue des deux mondes* in May 1874 and his report fired the imaginations of the French, intent on imperial expansion after being defeated in a humiliating manner by the Prussians in the Franco-Prussian War of 1870–71.

Like many Victorian advocates of climatic engineering, Roudaire relied heavily on select examples from classical literature to argue that the climate had been – and could further be – changed. With the ongoing deforestation of the hills surrounding the lake, he believed the once large inland sea had evaporated into smaller seasonal lakes. The ancient Greeks knew this lake as Tritonis. The lake once flowed freely into the sea when in the past the surrounding area received more rain. Arab traditions held the romantic notion that many of the oases of the Sahara were once seaports. The remains of this inland sea intrigued and repelled simultaneously. The undulating valleys proved fatal to many travellers. The unwary would often fall into soft sand fills and then disappear below, into ancient underground aquifers. Camels, laden with goods, and even the lone walker could disappear in a moment, never to return. Into this region Roudaire brought a team of surveyors to determine the feasibility of restoring the Sahara to its ancient condition.[22] He proposed that the chain of lakes should be filled with water from the Mediterranean Sea and called for a channel to be cut from the Gulf of Gabes into the depressions and thus refill a massive area of the central north Sahara.

Imperialistic French geographers gave lavish descriptions of the ancient economy, which produced 'gold dust, ivory, gums and ostrich feathers', to support grandiose claims for reviving ancient Roman climates through reforestation or schemes to create inland seas.[23] Roudaire's ideas received attention, but scientists and public intellectuals – as they did with many other schemes to change the climate by creating lakes – pointed out many of the idea's flaws. Quite obviously, seas do not magically create precipitation. The Dead Sea is surrounded by desert and some of the driest deserts in the world are located near coastlines, for example, in Namibia and Australia. The lake could possibly change the climate of the surrounding area, creating an oasis, but it would have no influence on regional climate more broadly.

The British trader Donald MacKenzie argued for a similar project in *The Flooding of the Sahara*. He proposed that a great inland sea could be formed by cutting through the sands in the north and letting the water pour into a central depression. The plan, he argued, was not as unlikely as it seemed on first impression. It would merely restore what had once been. Like empire foresters before him, he appealed to classical sources for justification. When Jason, sailing on the *Argo*, was blown off course and into an unknown bay, he offered sacrifices to Triton, son of Poseidon, far inland where the sands of the Sahara now blow. Scylax of Caryanda (late fifth to early sixth century BCE) wrote that 'this region, which is occupied by Libyans, is the most magnificent and fertile; it abounds in fine cattle, and its inhabitants are most beautiful and wealthy.'[24] The new inland sea would restore the same moist atmosphere that enabled Carthage to grow and sell its surplus grain to Rome, indirectly feeding the armies that built the Roman Empire.

The reading public found this project fascinating, much like other schemes to flood deserts. Public sentiment fanned the flames of ideas that garnered little scientific support.[25] The *New York Times* argued that the evaporation from this new lake would supply the 'aqueous vapor' around the edges of the lake and thus improve climate.[26] The American geographer George W. Plympton (1827–1907) adamantly defended the project to flood the Sahara. Sceptics in American journals had claimed that the creation of a new African sea would radically reduce the temperature of Europe for the worse. They argued that when the sea had previously covered the Sahara, Europe suffered a glacial age. They also claimed that commerce throughout the Mediterranean would suffer from lower water levels as the sea poured into the Saharan depressions. Plympton quoted one author who claimed that 'So much water drawn from the present oceans, may, by lessening the depths of the harbours of the world, produce serious and wide-spread inconvenience.' But all such fears, he felt, were 'utterly groundless'.[27] That is because most of the desert is in fact above sea level, and the water brought in by a new channel would still be relatively shallow. While creating a new inland sea, the project would add benefits without causing destructive climate change, nor would it significantly reduce the level of the sea in the Mediterranean ports.

Roudaire's reports so inspired Jules Verne (1828–1905), the famous novelist of *Twenty Thousand Leagues under the Sea*, that he wrote a novel, *The Invasion of the Sea* (1905), based on the prospect of flooding the Sahara. He stirred up imagined pride at this still-to-be-done achievement: 'A hundred years after the French flag was raised over the Kasbah in Algiers, we will finally see our French fleet sailing over the Sahara Sea, bringing supplies to our desert outposts.'[28]

Verne wrote his proposal for flooding the Sahara in the triumphant, optimistic spirit of late imperial engineering. In 1904 an American project had begun to carve out the Panama Canal that connects the Pacific and the Atlantic oceans. In 1869 the French designed and oversaw the construction of the Suez Canal, which linked the Mediterranean (and thus the Atlantic Ocean) with the Indian Ocean. Verne's imagination foresaw an inland sea that would allow shipping right into the heart of Africa.

Verne believed not only that flooding the Sahara would mean a ship could sail from London straight to Timbuktu, but that doing so could improve the climate. He wrote,

> In the first place, the climate of Algeria and Tunisia would be appreciably improved. Under the influence of southerly winds, the clouds formed by vapor from the new sea would bring beneficial rain to the whole region and increase its agricultural yield . . . After these physical improvements, who knows what commercial gains might ensue once this region was transformed by the hand of man?[29]

Verne's views reflected those of other foresters and explorers who advocated for the manipulation of the Saharan landscape.

Yet fear lurked behind Verne's optimism. He acknowledged that many critics had worried that the proposed sea would only send rain south and east rather than north, as French imperial advocates wanted, or that the waters would heat the atmosphere and breed mosquitoes and pestilence. Oddly out of sync with the optimism of his novel, Verne also worried aloud about the attempt to restore Eden in the Sahara. The Sahara was a vast solitude, 'that held a hint of something – no one knew what – but definitely mysterious'. The desert was seen as 'an ever-present but invisible danger – the feeling of an undefined threat, something comparable to the vague anxiety that precedes all atmospheric cataclysms'.[30] Other Victorian figures worried that the project could cool the air towards Europe and plunge the region into a new ice age, essentially destroying European civilization. This is an example of the fear of destructive climate change that could end life, as it was known at the time.

The Kalahari

The eccentric South African geologist Ernest H. L. Schwarz (1873–1928) had plans to abolish the Kalahari altogether by flooding it with waters from central Africa. Schwarz had emigrated from Britain to South Africa in 1895 to make his fortune from gold and to contribute to the geological study of southern Africa. He then found employment at the Geological Commission of the Cape of Good Hope, where he worked for ten years before taking up South Africa's first professorship in geology at Rhodes University in Grahamstown, located in the eastern Cape. His years working for the commission in the dry Cape Colony and his experience working with farmers in the Karoo seem to have given him a negative view of arid lands. He argued that because South Africa was a high table of land that fell off at the edges to meet the ocean, and because water from all sides of the vast interior cut channels through this elevated bowl and escaped to the sea, South Africans must 'turn off the taps' to keep the country from drying out. Because the evaporation of water is three times the amount of rainfall, the dry air sucks all moisture from the soil, and South Africa was at risk of complete desertification, Schwarz argued.[31] The region, he warned, was 'being wrung dry'.[32]

It was not always so, he argued. Modern-day Botswana (then Bechuanaland), he believed, once boasted large inland lakes that by 1820 had largely dried up. David Livingstone, the stern Scottish missionary, first proposed this idea in the mid-1800s.[33] These inland lakes, known today as the Makgadikgadi Pans, supposedly once evaporated vast quantities of water that then joined forces with the moist sea-borne air from the Atlantic and Indian Ocean. This dual action produced substantial cloud formations that resulted in plentiful rain. In the Karoo, now a vast semi-arid zone in the southwest, hippopotami and rhinoceros once roamed where prickly pear – an exotic, invasive desert plant – thrived, to the chagrin of white settlers.[34] That is why, Schwarz argued, the human population of southern Africa hugs the coast where water is more plentiful, and why the interior is unnaturally dry and undeveloped, and remained a 'wasteland' in the eyes of European settlers.[35]

National development of the interior remained a priority to the new white nation of South Africa at the establishment of the union in 1910. The country's variable climate, which was prone to droughts as well as downpours, caused anxiety among white farmers and scientists and raised the possibility that the country was progressively drying.[36] Plans to increase rainfall and conserve water received prominent public and government attention. The Cape Colony government instigated early reforms to protect trees for climatic purposes in the late 1600s under Dutch rule, but the Dutch spent as little money as possible on the colony, so tree planting was kept to a minimum. In the second half of the nineteenth century, the British colonial government, who took over the Cape from the Dutch during the Napoleonic Wars, tried to actively reshape the climate by planting trees throughout the colony, even in deserts. A significant portion of trees died or grew stunted due to prolonged droughts.[37] Dams often proved impractical because the mountains in South Africa were simply too low to produce enough snow to collect precipitation and then produce water run-off at a measured and sustainable pace. The reservoirs would thus hardly fill, and easily go dry, gutting the plans for wide irrigation. Because the shape of the high plateaus ran downward towards the coasts and quickly drained, the water that was diverted from higher elevations into the lowlands also tended to drain all too quickly.

Some water projects succeeded. Natural springs, and the run-off from the condensation of air that descended the face of Table Mountain, provided the city of Cape Town with what the Indigenous Khoisan had called *Camissa*, or sweet water. Canals and tunnels constructed under the ground in the nineteenth century distributed the water down the slope of the mountain and throughout the town. With a growing population, the Cape Colony's first prime minister, John Molteno (1815–1886), hired John Gamble (1842–1889) as the colony's chief hydraulic engineer to remedy the shortage of water and create the Molteno Dam. This project worked well and is still in use today, watering the Company Garden near the old Parliament House.

Armed with the Irrigation Act of 1877, Gamble began an attempt in the Cape Colony to loan money to farmers to construct water-storage projects. These efforts in South Africa mirrored the Desert Land Act, passed in the same year in the United States, to reclaim and improve arid land areas, raising the amount of land claims from 65 to 260 hectares (160–640 ac). Both had in mind the colonial movement of white settlers onto 'marginal' lands that could be made productive through irrigation.[38] In South Africa this and later schemes never progressed far because the conservative farmers distrusted taking loans at interest and worried that British administrators could strip Afrikaans farmers of their land if they fell into arrears.[39] Many farmers who did apply were rejected because they were already too indebted. Gamble opposed many large-scale projects he feared were white elephants.

Unlike Gamble, Schwarz was not afraid of grandiose plans because it was not his job to pay for them or run them. He proposed a bold scheme to bring the Makgadikgadi Pans back to life by turning inward the outflowing water that ran off the African shelf into the sea. Dams and canals could harness the flow of the Chobe and Kunene rivers to water a new and lovely garden in the heart of the country. Many geologists agreed that the Kalahari was once a lake, and the vast dry basin today was a bed remnant of a once vast ancient body of water. Why did it disappear? Schwarz suggested that the water that once collected in this ancient lake did what all water does – it carved a channel through the banks of the lake and out into the sea. This channel became the Zambezi river, that great and mighty drain of central East Africa, the fourth largest river in Africa and the single largest river to pour into the Indian Ocean. But it could be made to follow its old course, he suggested – with dynamite, engineering and dams on the Kunene and Chobe rivers where they converged north of the Zambezi river – and thus bring the ancient Kalahari lakebed back to life.

This bold proposal shocked officials in the Union of South Africa and kindled a public debate that ran until the late 1940s. The plan had a seemingly intuitive logic that appealed to lay readers and some politicians. Schwarz challenged the public and government officials to envision all the water that poured over the Victoria Falls – the largest and mightiest waterfall of its type in the world – spreading into the very heart of the desert of South Africa. This mighty feat, made possible by the luck of the landscape, would allow a mere turn of a river to be dammed, and, instead of escaping eastward, the river would pour into its old bed, draining the wet highlands of the north to the south. In this watery sphere, white settlers could navigate a huge body of inland water, irrigate fields and, through the evaporation of water into rain, support again the southern cone of Africa, transforming it into a new Eden – the genesis of life and ecological diversity that it had once been: a paradise regained.[40]

Schwarz argued that doing nothing would cause South Africa to slowly dry out and waste away economically. Conserving the water in the Kalahari would increase the rainfall that would in turn keep the rivers and lakes of the whole region sparkling with water and reverse the desertification process observed by Livingston when he first searched for the source of the Nile. If no action was taken, however, the Kalahari would continue to march south, east and west. Schwarz enthused that,

> When the gaps are blocked up and the old Kalahari lakes are once more there to supply the air of South Africa with moisture, the old central river of the Kalahari will once more flow. All down the course from Ngami to the Orange River, below the Falls, settlements will arise and agriculture, of the same nature and on the same scale as that in Egypt, will spring up. Yet it will be a better Egypt that will result from the occupation of this wilderness, for there will not be desert around. The Kalahari today is covered with bush and grass; when the lakes and rivers are reconstituted, the vegetation will be that of the Soudan; rich pasturage and forest.[41]

There will follow a 'clean sweep of vegetation' that will supply the stock farms and increase the wealth of the country. Boldly, he announced that 'Every acre of land will become habitable.'[42]

This 'thirstland redemption' would be better than irrigation, which waters only a few strips of land, leaving the rest as desert. Schwarz reminded the public that enormous sums of capital from rich investors poured into questionable irrigation schemes that had a high rate of failure. With the Kalahari project, 'everyone in South Africa, whether he wants it or not, will receive the additional rain, will see his land rendered more fertile, and all his difficulties from drought, famine and pestilence disappear.'[43]

Schwarz pointed out that North Africa once boasted emporiums of grain and wine that fed the ancient world along the shores of the Mediterranean. The Phoenicians, the Greeks and the Romans all fed off these rich and well-watered lands. What was North Africa now? A desert. The ruins of these once vast cities litter the coastline. The Sahara has taken in the entire hinterlands, and the desert reaches to the very sea. So, too, with the Kalahari. Schwarz predicted it would grow outward and become the new Sahara of the south. The cities of South Africa would crumble into ruins like those in Algeria and Libya and the countryside would boast only dunes and exposed sun-baked rock. Only one thing stood between desertification and redemption – the construction of dams to remake the ancient Kalahari lake. This engineering project would create a new inland sea of fresh water that would turn a drying South Africa back to the Eden it had been three hundred years previously.[44]

Scientists and public officials scoffed. In 1914 a committee appointed by the senate of the new Union of South Africa, and then another in 1920, the Drought Investigation Commission, did not find evidence of lessoning rainfall or a drying out of South Africa.[45] They concluded that the weather was variable, that South Africa was naturally semi-arid, and much of the country's land had been affected by bad land-management practices. Still, they argued, desertification was not advancing. The Kalahari would not become the Sahara of the south.

Public outcry about declining rainfall came from white farmers and the press three decades later. Another committee in 1947, staffed by officers of the department of agriculture, convened to consider the question of desertification. One of the officers, C.E.M. Tidmarsh, felt that the actions of man had indeed changed the environment for the worse. Alarmed by Tidmarsh's suggestion, the Desert Encroachment Committee concluded in 1951 that 'during the period of [the] white man's occupation of South Africa very considerable changes have taken place in the composition of the vegetation of different parts of the country,' and that these changes had been caused 'primarily from grazing, and the destruction of natural grasses'.[46] The natural vegetation slowed water in its course to the sea, and denuding the land of this vegetation threatened the topsoil through erosion.[47] Nonetheless, they concluded, in a damning indictment of alarmists, 'There is no proof, however, of any permanent diminution, and periods of plentiful rains may be confidently expected in the future.'[48] In light of this conclusion, the committee hoped to bury once and for all Schwarz's plan to flood the Kalahari. His plans, they concluded, 'merit[ed] no further investigation'.[49]

Meredith McKittrick, an environmental historian who focuses on southern Africa, has investigated the longevity of these ideas. She notes, 'Schwarz's scheme was immensely popular among the white South African public, despite the fact that three government investigations in as many decades declared it impractical.'[50] The endurance of Schwarz's ideas reflected the popularity of climatic engineering in the press and naive economic boosters. The history provides insights into how the public consciousness can be swayed by bold ideas presented in a simple format. It also should warn society today about the dangers of embarking on geo-engineering for global warming without extensive scientific, political and public debate.

The Battle for the Sahel

The exploration of and advancement of research on the Sahara informed thinking about the Sahel, an arid region that hugs the southern boundaries of the Sahara and starts in the west in Senegal and Mauritania and runs all the way east to Sudan and Eritrea through Mali, Niger, northern Nigeria and Chad. Unlike the sandy dunes of the Saharan interior, much of the Sahel features expansive savanna landscapes where native grasses grow amid acacia and other hardy trees that dot the landscape. The Sahel supports extensive pastoral and agricultural populations of people that are vulnerable to the droughts that periodically occurred.

French West Africa stretched all the way from the Mediterranean Sea down to Gabon in central Africa, some 4,000 kilometres (2,485 mi.) from north to south. French foresters in French West Africa, known to the French as L'Afrique Occidentale Française (AOF), raised the alarm in the late 1800s and early 1900s that deforestation was making West Africa drier, much like imagined deforestation had caused Algeria's climatic decline from breadbasket to desert. A French military official summarized the view of other officials and scientists by asking 'is it not natural to extend to the Sahara [and Sahel] the conclusions taken from the analysis of Algeria and neighbouring countries?'[51]

Foresters identified a similar set of problems which occurred across the Sahel. European foresters noted with alarm the use of fire for the management of grasses. They feared fire, whereas Indigenous people knew that fire helped to regenerate the grasslands, allowing for healthier, more palatable grasses to grow anew and hold at bay weeds and less desirable vegetation. A series of forestry commissions, held in 1907 and again in the 1920s, enquired into the condition of West African forests. Diana Davis, a noted environmental historian and geographer of the Middle East, writes that, 'By the late 1920s . . . the debates [over climate change] were largely over in Africa and a consensus was forming that anthropogenic desiccation was the main problem.'[52] The global severity of droughts in the 1930s sparked a flurry of alarming publications by British and French scientists, including Auguste Chevalier (1873–1956), André Aubréville (1897–1982) and Edward Percy Stebbing (1872–1960). Despite agreement among foresters and botanists about the need to protect forests, officials in French West Africa only got around to passing a Forestry Code in 1935, some thirty years after the French developed the Algerian Forestry Code (1902) and almost twenty since the passing of similar legislation in Morocco in 1917.

A former chief forester of British India, Edward Percy Stebbing, sounded the alarm about the common practice of shifting cultivation, a practice where people would burn vegetation in order to plant crops before moving on to a new area in the following years. Like foresters in India, who tried to suppress shifting cultivation because it contradicted European ideas of forest management, he believed that this practice turned West African deciduous forests into deserts. Stebbing first learned about forestry in West and North Africa from French sources. The French had a vast North African empire that provided the laboratory for French foresters, geologists, geographers and a range of other scientists to inform and debate theories about climate change. Stebbing became a passionate advocate of afforestation (or in his mind, reafforestation) because of his travel to West Africa.

Stebbing wrote an account of this shifting-cultivation practice in 'The Encroaching Sahara', a paper read at a meeting of the Royal Geographical Society in March 1935 and published in the society's *Geographical Journal* in June, where he argued that the widespread practice put all of Africa at risk.[53] Stebbing did not want deserts to be viewed as natural, eternal formations – geological features like mountain ranges or the vast ocean. Humans, he said, created the Sahara and other deserts. Comparing the observation of the great expeditions across the interior of the Sahara in the nineteenth century with conditions as they existed in the mid-twentieth century gave evidence of this. Henri Duveyrier (1840–1892) explored the Sahara in the 1850s on an expedition from Morocco to Tunisia, and in his journals he recorded seeing scraps of brush, grass and trees – even forests – that by the 1930s wholly disappeared.[54] In 1914 a German officer and forester, Baron Geyr von Schweppenburg (1886–1974), explored further, particularly in southwest Algeria and the Sudan. Publishing his results in 1920, he revealed that the Sahara contained petrified trees in sandstone beds, colourful murals produced by civilizations that had hunted a wide range of animals, and old riverbeds with washed gullies and eroded rock where water once flowed regularly and even flooded. Stebbing concluded that these 'vast deep-cut valleys' prove that 'mighty rivers' once flowed throughout the Sahara, and that this green land once extended 'from the Saharan Atlas to the bend of the Niger'.[55]

Earlier, when visiting the Gold Coast (modern-day Ghana), he noted in his diary, on 19 February 1934:

> There is no doubt that formerly a vast mixed deciduous forest occupied all of this region . . . Evidence is to be seen of this in the present condition of the savannah lands. Clumps of forest still remain, scattered old isolated trees are to be seen . . . the more one studies this country the more evident it becomes that the savannah lands are the degraded conditions of a region which was once covered with high forests.[56]

Evidence of the once mighty forests remained in the lone tree, grove or, in southern Nigeria, substantial patches of isolated forest. Therefore, it is wrong, he argued, to do as so many foresters and botanists have done – to pronounce the savanna as a biological community and as a type, and to either preserve it or allow it to expand. This 'acceptance of degraded mixed deciduous forest as savannah' detracted from understanding the true potential of the region.[57] Staying at a forest rest house in Owo, in the Ondo Province of southwest Nigeria, he wrote on 13 March that he saw a glimpse of what once existed, and should be again:

> [A] distant range . . . hazy . . . appeared as a sea of forest, but with dark-topped tree crowns standing up at intervals from the bush, or so-called savannah, thus indicating only too clearly that they were the last remnants of the old moist mixed deciduous high forest.[58]

If the savanna was not fought and forest reserves established, 'the ultimate result' would be desert. The future would be bleak if action was not taken. He argued that northern Nigeria increasingly looked more like the arid lands in 'the Punjab, Sind and Baluchistan in northwest India' than southern Nigeria.[59] Stebbing advocated a bold plan to solve the problem by planting a vast belt of forest around the southern edge of the Sahara, to hold back its advance. This suggestion alone may be one of the largest environmental proposals of its type ever advocated. However, the onset of the Second World War distracted officials from seriously considering the suggestion.

It was not only Africa at risk. Stebbing argued that desertification threatened the United States and Central Asia as well. With the sudden and dramatic appearance of the Dust Bowl in the United States, 'it [became] easier to read the writing on the wall.'[60] Soil, and how humans treated it, was the common thread that tied these deserts together. Human abuse of the soil changed the landscape so slowly as to be almost unnoticeable. Farmers had laid waste to Inner Mongolia through bad soil management, by forcing agriculture on a soil far too thin to support it. Along with overgrazing and the loss of topsoil, nothing could stop the advance of the desert as it displaced populations while stretching further south into the heart of China.

Although Stebbing argued strenuously against most geologists and other scientists who saw deserts as being natural and ancient, he was not a lone voice crying out in a vacuum. The Royal African Society, in December 1937, convened a meeting chaired by the parliamentary under-secretary of state for the colonies, the Marquis of Dufferin, to examine the problem of desertification. The speeches laid much of the blame for human-induced deserts on agricultural practice and advocated improved treatment of the soil.[61]

The climax of efforts to protect Earth from devastating climate change and desertification occurred with the 1947 Empire Forestry Conference in London.[62] This conference pulled together experts from around the world, including representatives of other European powers, and the newly formed Food and Agricultural Organization of the United Nations. Most of the delegates at the conference, which was sponsored by the British crown, had no idea that the British Empire stood on the edge of imminent collapse. At this conference, a breathtaking planning session ensued. The delegates proposed – along the lines suggested by Stebbing – not only to plant a belt of trees around the southern border of the Sahara, but to reforest a significant portion of the Middle East, plant interior portions of Australia with forests, protect the topsoil of the Himalayas from erosion, and save the remaining tropical forest areas of Africa and Southeast Asia. A supreme sense of confidence prevailed among these imperial officials who, fresh from the planning authority and resource scarcity of the Second World War, laid out plans that would regenerate much of Earth's land surface, advance development and protect climate. All these radical green visions collapsed with decolonization.

Green Prophet

A unique climate prophet arose in the mid-twentieth century who also shared an imperial past as a scientific forester and captured a loyal segment of public opinion for decades after the Second World War. Richard St Barbe Baker (1889–1982), an English botanist and writer, spent decades warning the world about the threat of desertification. He based his alarming rhetoric on the increasingly outdated theory of climatic botany. He promoted a profoundly mystical and romantic view of nature, and of trees in particular. His popular books were autobiographies that focused on his own daring role in the adventure of building an important international movement – the Men of the Trees – which promoted the planting of trees around the globe. His role in establishing this organization enabled him to meet prominent world figures. He wrote innumerable newspaper articles, and his many books sold well and are still found in used bookstores throughout the English-speaking world. Though no biography of him has been written, it is unlikely that his stories about his own central role in world events adhered strictly to the truth.

A powerful sense of desperation drove Baker to conclude that something had to be done to fight back against the encroaching deserts. He appealed to his audiences with grandiose and alarming claims from around the world, such as that during the Dust Bowl in the United States the topsoil that covered 82,800 square kilometres (32,000 sq. mi.) of land across seven states literally rose into the air and blew into the Atlantic, destroying the livelihood of thousands and threatening to open a new desert in the heartland of America.[63] In Australia he pointed to the destruction of forests that induced droughts and threatened to turn the eastern coast, a strip of verdure forest and farming land, into the same desert as the red-hot interior. In New Zealand the emerald forests that once covered the mountains and hills had been cleared for grazing, rinsing the topsoil into the sea, and leaving nothing behind but gullies and bare rocks. Latin America, in particular Argentina and Brazil, lost valuable topsoil every year due to deforestation. Similarly, in China the Yellow river emptied topsoil into the sea, adding to the spread of desert.[64]

For Baker, the Sahara stood as the great blot of aridity that threatened the whole of Africa and the Middle East. He saw at first hand the remains of cities and agricultural systems in present-day deserts where nothing would grow, such as the ruins of Djemila in Algeria, an ancient Roman city, surrounded by dusty dry hills, which once supported a large colony of farmers and traders. Baker believed, as did Stebbing, that savanna was degraded forest land on its way to complete desertification. The forests of the Sahara, Baker argued, once rich with rivers and pools of water running through them, were destroyed by invading Arabs who brought with them one hundred goats each. This massive invasion of animals ravaged the land, not only of grass but of saplings, killing the regenerative power of the forests.[65] The fact that cave drawings of chariot races exist where now only sand blows was proof to Baker of the once abundant Sahara forests.

In response to this threat, he launched in the 1950s a dramatic expedition to cross the Sahara by car, which he dubbed the 'Sahara Challenge'. The UK's *Daily Telegraph* provided seed money and potent coverage of the adventure. Beginning in London, ferrying to Europe, driving south to the Mediterranean through Spain, ferrying to North Africa and then driving south across the Sahara interior all the way to sub-Saharan Africa, he managed to create a substantial media event and laid the foundation for future sales of his book, also titled the *Sahara Challenge*.

While driving south across the central Spanish plateau he observed an increasingly arid landscape. This he blamed on the use by farmers of chemical fertilizers. The observation shows the extent to which the small and burgeoning organic farming movement had influenced his thinking. It frightened him to see the farmland abused to such an extent that it resulted in 'rock protruding through the soil' over much of Spain, with the irony of 'conspicuous advertisements for chemical fertilizers' prominent on the tight hairpin turns of the road, 'as if accumulated forest humus of centuries could be replaced with something magic from a bag!'[66] He saw a clear link between agriculture, deforestation and increasing desertification. These three things in conjunction made it 'reasonable to conclude that, for the greater part, the Sahara Desert was created by humans'.[67]

How did this process happen? First, the forest is cleared. Second, crops are sown. Non-organic farming methods do not return nutrients – that is, nutrients that took hundreds of years of forest cover to build up – to the humus. When the nutrients are consumed, the agricultural cover dies off or is abandoned because of lower yield. When this happens the ground is bare, with neither forest nor crop cover, and the forces of wind and rain begin the inevitable process of erosion. Further, without trees there is less recycling or storage of water. The surface soil dries out, streams dwindle, springs dry up and there is less rainfall. The soil then begins to blow away as great clouds of dust. 'The stored-up fertility of a million years may be lost in a single season.'[68] Worse follows. Without tree cover and the accompanying rain and dew, the temperature of the land rises, and clouds pass over without dropping any moisture. The region begins to warm, adding to global warming, and this in turn creates more favourable conditions for the deserts of the entire world to prevail.

Baker concurred with Stebbing that the savanna to the south of the Sahara was by no means a natural phenomenon but rather degraded forest in danger of slowly folding into the larger Sahara Desert. The slash-and-burn firing of the forest created a landscape of scattered trees until this too was burnt away. It was humans, according to Baker, not locusts, who destroyed the once verdant Sahara.[69]

Baker was convinced that this southern march of the Sahara had pushed refugees off their failing farms and had thus led to the collapse of whole civilizations in the past. He feared it would do so again. As he made his way through French West Africa, just to the north of Nigeria, he compared what he saw with Livingston's observations of the previous century. Livingston had supposedly seen Nyasaland (Malawi) covered with great rainforests. Since that time these had all but disappeared, the rainforests degraded into savanna and finally to drifting sand. The dramatic change had occurred over his own lifetime. He now had to 'trudge through sand wastes which had been my forest haunts when I had been in Africa thirty years ago'.[70]

Baker foresaw Lake Chad drying out and filling with silt and sand. By comparing century-old maps with those he had in his possession in 1952, he believed that the once vast lake had lost seven-eighths of its size, with sand filling its former bed. In his lifetime the lake had shrunk to a miniature version of its former self. In Sapoba, the forest station where he had once served as a British forest officer in northern Nigeria, sand dunes had recently pushed right up against dense rainforests, showing little transition or buffer from forest to savanna to desert. He mused with a fellow empire forester that if only the British could get the cooperation of the French to ban grazing over the savanna, much of the forest would return, and a southern belt around the base of the desert could contain the violence of the desert assault with an emerald wall of protection.[71]

Baker rejected the idea of secular, that is, natural, climate change. Only wistful thinkers who loosely talk 'of wet and dry cycles and would like to think that the present period through which we are passing is just a long dry cycle' asserted that Earth went through cycles of climate change.[72] Everywhere on his expedition he heard the same complaints: of diminishing rainfall and drying countries. As he drove to Kenya, from west to east on a 14,500-kilometre (9,000 mi.) journey that skirted the southern edge of the Sahara, he could reveal to his fellow British settlers in that country that the Sahara was racing southward. They were 'alarmed when they realized that their chosen home might soon be enveloped', and it became apparent that 'the drying up conditions that they themselves had observed were ominous portents.'[73]

Baker believed not only that deserts were human-made, but that theorists who postulated elaborate meteorological reasons for climate change simply abrogated the duty of humans to stop the damage. The theories proposed let humans off the hook for responsibility. Baker's answer to this was the reforestation of mixed species that were largely, though not wholly, composed of indigenous kinds. He opposed planting mass skeletal-like plantations of eucalypts, which, he believed, only dried out the land further because the genera pumped water out into the air but held little moisture in the soil. He called these artificial plantations 'the land where the rainbow ends'.

> It was a relief to come to one of the rare remaining indigenous forests and over it hung the most vivid rainbow I have ever seen. It looked as if my road went straight through the middle of the western arch, but it always kept ahead of me seeming to move as I moved. At last, it stood still and I drove right into a light shower of rain, and at that point the rainbow ended. I had come to the end of the indigenous forest and had entered the eucalyptus world.[74]

Science, Baker felt, often did as much harm as good. Revisiting a forest research station that he had once run as a ranger, he was appalled at an American scientist who showed no interest in native forests or the larger picture of desertification. The new staff member announced, bluntly, 'We are pure scientists.' Baker wrote,

> It sounded like a slogan; all they needed was a banner with a scientific heraldic device. I thought to myself, 'pure science be damned.' Here is a state of emergency, a state of war. The great

> Sahara desert is invading Africa along a two thousand mile front at the rate of thirty miles a year in some places.[75]

Perhaps this disdain for science explains why Baker turned to literature to advance his ideas, most particularly in his novellas and short stories, where his imagination could conjure up an ideal future for humans living in harmony with nature. In *Kamiti: A Forester's Dream*, Baker tells the fantastical story of an African boy in Kenya, who, like *The Jungle Book*'s Mowgli, grows up to be a forester trained in Britain's imperial forestry tradition. In this novel, Baker let loose all his highest-flying fantasies about planting trees from the ground up and employing the Indigenous people – including the Mau Mau rebels who fought British rule from 1952 to 1960 – to push back the Sahara Desert. In this environmental vision he illustrated how the peoples of the world could work together for conservation and establish global peace. Science, faith, fantasy and Baha'i doctrine, of which Baker was a devotee, combined to create an African warrior facing off against the greatest threat to mankind: the desertification of the world.

In *Kamiti*, the protagonist Kamiti learns to love the lore of plants and animals in the forests of Kenya, including 'healing balms and bushes' that restore his beloved tribe to health.[76] After his father dies, he meditates in the forest and falls asleep. As he dreams, a star, larger than the others, moves closer. 'Your people need you,' the star says.

> Strange trees have been planted that are thirsty for water and dry up the land, and the desert increases. You, my Kamiti, will go to Old England, and work in the schools and learn the story of the trees so that you may return . . . master of the trees . . . and serve your people.

Kamiti wakes up, struggles to his feet and realizes that he is now a medicine man, a doctor of trees, whose mission is to restore the skin of the sick Earth. Kamiti prophesies, 'The tired Earth will live again and become a garden and a paradise. This is the hour of the coming together of the sons of men . . .'[77]

Kamiti's training begins in England. In the novel, forestry students in England had donated money to a scholarship fund to support an African student and so, on the generosity of his future colleagues, Kamiti goes north on a scholarship, 'from the Garden of Africa to the Garden of England'.[78] He attends New College, Oxford, and the School of Forestry that trained young men to be foresters in the British Empire. The students came from the white dominions, but also from the colonies, the newly independent nations and even China. Here Kamiti finds 'quite a brotherhood of foresters'. He studies all the life sciences, including chemistry, botany and ecology, along with the principles of silviculture. He learns that a tree has 'a capacity for infinite growth', as did forests – trees belonged everywhere and had been destined to thrive in all parts of the Earth. But Earth's forest treasures were all threatened, as was the 'red gold' of Nigerian mahogany where even forestry protections 'would be broken and the Kingdom of the Giants would pass and soon the Sahara would advance towards the coast'.[79]

Transferring his training to France, Kamiti notices in Europe 'the perfect balance between farm and forest', including the use of horses instead of tractors on the land.[80] Man has 'tortured the soil' first by deforestation and then 'with chemicals and fast-moving iron ploughs which disintegrate the water in it and disturb the electro-magnetic force' that protects the dance of life.[81] In Nancy, formerly the Duchy of Lorraine, in northeast France, his fellow empire foresters prepare to tackle the environmental ills of the world. All other issues fade in importance. What could be more vital than saving the Earth? 'How petty were our politics in the face of the threat of the desert. Even all the atomic frightfulness fades before the speed at which the deserts of the world are undermining life.' Even an atomic war could not equal the threat.[82]

Kamiti's life calling comes after his time in Nancy. He realizes he must lead a 'Green Front' to attack and push back the deserts, then 'reclaim the denuded Earth and make it fruitful again'.[83] In another dream, Kamiti finds himself flying through space in a ship, 'gliding past the stars and looking down on the circling galaxies of spheres'. Soon he came to a green star. The star was in reality a planet. Surveying the planet below he could see that Man had never set foot here. 'Trees alone ruled supreme, with the forest creatures each playing their part in the life of the forest.'[84] All the animals and plants worked together in harmony. It proved to be 'my model forest . . . what the foresters of Nancy called the forest climax'. Waking, he drank clear water from a spring untouched by metal pipes or a dam, 'charged with magnetic force' and ready for his mission to change the world.

Anointed with his mystic mission, Kamiti recruits ten men, who each recruit another ten men. Soon he has a small army of amateur foresters planting trees. According to Kamiti, *Casuarina* and *Acacia*, Australian trees, would heal the wounds on the Earth's skin. *Casuarina* draws nitrogen from the air and enriches soil that has been drained of nutrients. The governor of British Kenya sends Kamiti a letter from Government House in Nairobi to offer his help. Kamiti asks for 20,000 Mau Mau prisoners, imprisoned for violent insurrection, whose labour could be focused on saving the planet, which would transform them into soldiers of the green front. After some hesitation, the governor agrees. The new army of recruits marches along the south of the Sahara, planting nurseries for saplings and replanting these in due time on farms and desert areas. Unfarmed areas become forests. Farmed areas of 'mixed farming', along the lines of organic farming, balanced woodland and cultivation. The desert advance is halted and the United Nations asks Kamiti to lead his green front out of Africa to the rest of the world. World peace dawned at last on a new green planet.[85]

Kamiti represented Baker's vision of the world. Baker's international society Men of the Trees attempted just this vision, minus the dramatic simplification. Baker had a lovable and charismatic personality, but his writing is marred by so much self-promotion that it is difficult to know when the author is merely exaggerating his impact on global events, or if he wholly fabricated events. The society he founded, however, the Men of the Trees, certainly did make headlines around the world, and its adherents planted millions of trees. As with organic farming, mysticism helped communicate science to a broader public, and Baker shared, like many foresters in the British Empire, a doctrine of holism that fed from and into the broader environmental movement of the 1930s to the 1970s. Baker's ideas on climate change are important not because they are scientific, or correct, but because he reflected the ideas of many in conservationism that swiftly transformed into the modern environmental movement. Baker was a mystic, and he took strands of mysticism – those that resided in the romantic strain of conservation – to serve as conduits in the development of the modern environmental movement, much as did his fellow travellers in the organic farming movement. Many of his ideas, formed in the 1920s and '30s, and expressed in his writings of the 1950s and '60s, still have resonance today. Baker is interesting because he was one of the last foresters to publicly advocate for the idea that forests influence precipitation after the Second World War. Though he fought valiantly, his views could not succeed against the headwinds of meteorological and hydrological research, the subject of the following chapter, which summarily dismissed the idea that forests influenced precipitation.

6

HOW DREAMS OF RECLAIMING DESERTS EVAPORATED

The Great Depression in the early 1930s coincided with severe droughts and dust storms that made it seem as if nature itself was wrathful. Rainfall in the American Midwest plummeted in 1931, less than two years after the crash of the New York stock market, and it did not recover until a decade later. Drought came to the forefront of public attention on 14 April 1935, when a colossal 1,600-kilometre-long (1,000 mi.) dust storm hit Texas, Oklahoma, Kansas and surrounding states. Robert Geiger, a journalist working for the Associated Press, coined the term 'Dust Bowl' to describe the storms battering the Great Plains. Newspapers carried pictures of the storms swallowing whole towns across the American flatland, from Texas to Oklahoma. Other regions of the world, such as Australia, also experienced severe droughts and dust storms due to decreased rainfall and increased temperatures. Red-coloured dust storms came in from the central desert to envelop Sydney and other cities in Australia from the 1930s to the mid-1940s.[1]

Desperate for rain, residents and experts in drought-stricken areas invoked religion, magic and science to pry open the heavens. Religious congregations prayed for rain. Newspapers published reports of Indigenous rain dances. When prayers and rituals failed to work, itinerant rainmakers tried to convince farmers and small communities to pay for chemicals, rockets and electromagnetic forces to do the trick.[2] Rainmakers made small fortunes preying on the desperate hope of drought-stricken communities.

People turned to religion and even quackery because meteorologists offered few tangible solutions. Willis Gregg (1880–1938), chief of the U.S. Weather Bureau, poured cold water on the idea that humans had the ability to manufacture rain on demand in a radio interview in Chicago in July 1934.[3] Gregg, like other meteorologists, thought the movement of large masses of moisture through the upper atmosphere determined when and where precipitation fell. The scale of magnitude – thousands of kilometres – meant humans could have little to no influence over where rain fell. Gregg believed weather modification by humans would take centuries, if it could ever happen.

Farmers did not want to wait another decade, let alone a century or more, for meteorologists to make it rain. When government meteorologists stonewalled the farmers' requests, a range of other scientists offered a glimmer of hope. Buoyed by public interest, physicists and foresters concocted schemes to increase rain in drought-struck areas. These ideas gained legitimacy on the back of hydrological theories and evidence which suggested that local evaporation and transpiration produced a large amount of precipitation. If correct, humans could increase rainfall by retaining water in lakes and soil. The more moisture saved, the more precipitation would occur.

John C. Jensen (1880–1957), professor of physics at Nebraska Wesleyan University, was one such scientist who proposed this idea. At the 1934 meeting of the American Meteorological Society, in Pittsburgh, Jensen called for a programme of impounding water in lakes and ponds across the Midwest to increase rain. He wanted to store 40 per cent of all precipitation in dams and ponds to increase the availability of moisture in the air.[4] Jensen suggested that the Midwest had the necessary ascending motions of air to generate more rain if extra moisture was added into the atmosphere: 'in Nebraska there are thunderstorms anyways: give them more vapor and they'll rain.'[5] He cited research from Australia suggesting that artificial reservoirs increased rain in the state of Victoria.[6] U.S. president Franklin D. Roosevelt's Civilian Conservation Corps (CCC), established in 1933, offered the labour required to pursue such a gargantuan task.

Jensen's paper raised the intriguing question of where moisture for precipitation comes from. He believed that local waters – evaporated water from lakes and the land – formed a significant proportion of the precipitation in that locale. With this logic, if you could keep lakes in a region full, then that region would reap the benefit of more rain. In Nebraska, where Jensen lived, this line of thinking could be traced as far back as the 1860s, when people advocated the idea that 'rain follows the plough.' Whereas Jensen focused on impounding water, an earlier view, advocated by Samuel Aughey Jr (1831–1912), a professor of natural sciences at the University of Nebraska, advocated that the Midwest's hard soils needed to be tilled so that water could penetrate and saturate them.[7] Like a sponge, the water in the soil would eventually evaporate into the atmosphere rather than running off into rivers and streams and thus out into the sea. Increasing moisture in the atmosphere would increase rain. Aughey's ideas poured fuel on pre-existing beliefs that tree planting could modify the harsh climate of the Great Plains. Government legislation had since the 1870s encouraged tree planting in the Midwest. The Timber Culture Act of 1873 gave American colonists 65 hectares (160 ac) of free land if they 'shall plant, protect and keep in a healthy, growing condition for ten years forty acres of timber, the trees thereon not being more than twelve feet apart'.[8]

Hope spread like tumbleweed across the Plains in the 1870s and '80s. Advertisements called arid Nebraska the 'Garden of the West' and claimed that the expansion of railways, with the homesteaded land surrounding them, was turning deserts into farms. Frederic Clements (1874–1945), an esteemed Nebraskan-born ecologist, later described the naive enthusiasm of the era: 'The belief was fostered in every conceivable manner by the ambitious young commonwealths and by all those with lands to sell, to the extent that university professors were able to find convincing proofs [that the climate was changing] and discover the causes.'[9] When drought hit the Midwest in the 1890s, these ideas started to lose favour. Nonetheless, adherents continued to believe that ploughs needed to go deeper into the soil and more trees needed to be planted. Advocates for ploughing made the point that, as of 1900, only 8 per cent of the land was cultivated. Who knew what would happen if more of the region could be put under the plough? An influx of new farmers, bolstered by good rains in the 1910s and '20s, led to a revival of hopes about improving the climate.[10] Like the crushing drought of the 1890s, the lack of precipitation in the early 1930s shattered these visions of agricultural plenitude.

Yet the idea that humans could influence rain remained well into the 1930s and beyond. Even Clements, who dismissed the idea that rain follows the plough as infeasible in practice, agreed that increasing the availability of local moisture could theoretically increase rainfall. His research suggested that increasing retention of moisture in vegetation or soil could possibly modify precipitation. He ran a study comparing transpiration and evaporation rates of native grasses compared to farmed grains.[11] His 1923 publication concluded that native grasses and grains showed no differences in moisture retention. Thus the replacement of native grass sward by fields of grain did not increase or decrease moisture availability. Clements's study did not challenge the idea that vegetation influenced moisture, and thus he was at times interpreted as supporting the belief that local moisture influences rain.[12]

Confusion reigned before the Second World War in part because scientists lacked clarity on the movement of moisture through the atmosphere. This led to a variety of conflicting interpretations. Key scientists in the fields of ecology, hydrology and forestry continued to support the idea that precipitation was heavily influenced by continental evaporation.[13] Government agencies and leading experts in hydrology and forestry made claims that at least implied that rain came from evaporation. A 1934 report from the U.S. National Resources Board noted, 'A considerable portion of the precipitation over any region is derived from local evaporation from that region. Only that part of the precipitation reflected as run-off in streams of the region draining into the ocean is derived from ocean sources.'[14] The best way to create rain would be to increase the availability of moisture in the atmosphere. All of this helped maintain popular beliefs that rainfall could be manipulated. Newspaper evidence from Australia showed that, through the 1930s, a majority of articles discussing forests and climate supported the idea that forests influenced rain directly.[15]

Meteorologists from the 1930s to the 1960s waged a continuous battle against the idea that humans could modify precipitation by conserving moisture in forests, rivers, lakes and the soil. Meteorologists had to convince the public that local precipitation came from moisture hundreds or even thousands of miles away. Along the way, they discovered that many deserts had higher levels of moisture in the atmosphere than some areas that receive abundant rain. They pointed to evidence accrued from nearly fifty years of weather balloons, the advent of flight and new theories of weather fronts and upper-level air movements.

Debating the Movement of Water and the Origins of Moisture

Where does the moisture in rain or snow come from? Is it derived from local sources or from far away? If a region's precipitation comes from water that evaporated from local sources, such as a lake in the desert, then it makes sense to ensure that water does not 'escape' to another region or the ocean in rivers and streams. If atmospheric moisture comes from the ocean, then planting forests or diverting rivers will be of little consequence to precipitation.

How water moved on the ground and in the atmosphere has long remained a mystery because of the difficulty of studying tiny water molecules. Scientists knew that water moves across vast distances and goes to incredible heights, but they lacked accurate ways of observing it. Even the advent of flight did not immediately solve this. A report in 1942 by American meteorologists noted: 'Since the water entering the atmosphere becomes a gas and is invisible, direct methods such as those in use for measuring precipitation or run-off in streams cannot be employed.'[16] The use of mass spectrometers in the 1950s and early 1960s finally allowed for the measurement of distinct oxygen and hydrogen isotopes within water. Even with these technologies, however, the complex dynamics of cloud microphysics bewildered efforts to understand – and manipulate – precipitation.

Humans have speculated on the origins and movement of Earth's waters for millennia, invoking religion, magic and science to explain why it rains and how to create or stop precipitation. The question of where rain comes from also led to others, such as why do the oceans not overflow despite the steady flow of rivers and rainfall into them? One must go to the early modern era (1500–1750) to find the seeds of contemporary thought.

The Renaissance craftsman and artist Bernard Palissy (1510–1589), known as 'Palissy the Potter' for his efforts to create porcelain like the Chinese, is credited with challenging the idea that the Earth's water came from underground caverns, a belief that some early modern naturalists had propounded. Palissy believed that God created the world as a dynamic system made in perfect balance. He wrote, 'God did not create these things to leave them idle . . . what is consumed naturally in it is renewed and reformed again.'[17] The Frenchman argued that rainfall provided the mechanism of 'renewal' of the Earth's water, which, according to the Bible, was 'all created in one day'. An unorthodox person and a bold thinker, Palissy eventually died a martyr in the Bastille, the infamous Paris prison, because he would not renounce his Protestant views in Catholic-dominated France.

The rise of formal scientific societies in Western Europe in the late 1600s and early 1700s helped usher in a new quantitative age of water research founded on empiricism, measurement and debate in the publications of learned institutions such as the Royal Society.[18] The French naturalist Pierre Perrault (1608–1680) published a key book, *On the Origin of Springs*, which further developed modern thinking on water. Perrault, like Palissy, suffered hardship in his life, and his extreme debt forced him to give up his prestigious position as Receiver General of Finances for Paris. In lieu of money, he turned to literature and the study of hydrology as a recompense.[19] Like Palissy, Perrault was not afraid to go against orthodox views. Perrault designed innovative experiments which showed that water only penetrated a few feet at most into soil; this offered proof that water did not seep deep into the soil into subterranean caverns. He also demonstrated that there was more than enough water that fell as rain or snow to feed all the rivers and streams. This further encouraged the idea that Earth's water cycled from air to land to sea and back again.

One of Britain's foremost naturalists, Edmund Halley (1656–1742), established key empirical and theoretical principles on evaporation that underpin much of modern hydrology and hydrometeorology. The idea that water vapours moved in the atmosphere gained acceptance among many naturalists after Halley published a series of findings beginning in 1687. Based on his measurement of evaporation – roughly 0.25 millimetres in a twelve-hour period – he extrapolated how much evaporation occurred in the Mediterranean Sea daily, and then he compared this figure with his estimate of streamflow based on the rivers flowing into the sea. He concluded that evaporation from the ocean proved entirely sufficient to supply the water for streams.[20] This explained why the ocean did not overflow, which would happen if rivers continued to flow into it (like water into a bath) without evaporation. Halley's method of using a pan to measure evaporation was similar to experiments conducted today, although there has been a significant amount of research since about how the size and location of water-catching pans influence the measurement of evaporation.[21]

If Earth's water moved in a cycle, where did it come from and why did it fall on some places and not others? The movement of moisture in the air proved difficult to measure and theories abounded about the movement of water vapour in the atmosphere. In 1864 George Perkins Marsh (1801–1882) described the difficulty of making any accurate measurement:

> The currents of the upper air are invisible, and they leave behind them no landmark to record their track. We know not whence they come, or whither they go. We have a certain rapidly increasing acquaintance with the laws of general atmospheric motion, but of the origins and limits, the beginning and end of that motion . . . we know nothing.[22]

Marsh also recognized what many mid-twentieth-century meteorologists and hydrologists would later say about schemes to make rain: water molecules moved long distances and thus any effort to plant trees or make a lake might have little or no impact on local precipitation: 'it is always probable that the evaporation drawn up by the atmosphere from a given river, or sea, or forest, or meadow, will be discharged by precipitation, not at or near the point where it rose, but at a distance of miles, leagues, or even degrees.'[23]

People offered various theories for how long water circulated in the air before falling as precipitation. In the 1860s, some, such as the engineer Baldwin Latham (1836–1917), believed that 'circulation is slow and gradual – so slow that the spherule of vapour now rising from the ocean may be years, or even ages, in returning to its native source.'[24] By the early twentieth century, meteorologists proposed a much shorter lifespan, around twelve days.[25] The inability to measure the upper atmosphere hindered scientists' ability to make informed conclusions about how, how far and how much moisture moved through it.[26] These ideas remained speculative until the development of tools to measure the upper atmosphere that came with the greater use of aeroplanes and balloons after the First World War and then mass spectrometers after the Second World War.

The View from the Ground: Hydrology and the Shift to Demand-Side Thinking

With uncertainties about how moisture moved through the atmosphere, scientists tried to make better sense of data collected on the Earth's surface. By understanding the movement of water once it reached the ground, scientists could better conserve it for human use while also possibly increasing moisture availability in the atmosphere. If scientists could not make rain, they could at least conserve water in rivers, soils and lakes. The field of hydrology emerged as a discipline in the 1930s because of the pressing interest at the time in water conservation, which had been heightened by droughts across the world, especially in the United States.[27] More than anything, hydrologists shaped policies for managing terrestrial water, but they also played a key role in the debate about the relationship between local moisture and precipitation.

Since the mid-nineteenth century, measurements had shown consistently that more precipitation fell on land than ran off into the ocean from rivers. This meant that a significant proportion of rain evaporated or transpired back into the atmosphere or seeped into the ground. In the 1930s and '40s, estimates of this 'lost' water ranged from 33 per cent to as high as 75 per cent.[28] Measuring the so-called lost water gave insight into whether moisture that fell as precipitation came from the ocean or from moisture derived from the land. Evidence suggested that the amount of precipitation coming from water vapour in the ocean varied from as high as two-thirds to as low as one-third.[29] The more water vapour that came from the oceans, the less that terrestrial sources in the forms of river and groundwater mattered for rainfall.

The patchiness of rainfall made it difficult enough determining how much rain fell across an area, let alone in trying to ascertain where that rain came from. Following water once it landed proved even more difficult given the tendency of water molecules to seep into soil, transpire through vegetation, evaporate from the ground, and flow into a myriad of rivulets, streams and rivers as it moved to the ocean.

Robert E. Horton (1875–1945), an American scientist considered by many to be the father of modern hydrology in the United States, published an article in 1931 titled 'The Field, Scope and Status of the Science of Hydrology' in the *Transactions of the American Geophysical Union*. It is often seen as the intellectual origins of professional hydrology in the United States, as well as elsewhere in the world. Horton called for the establishment of a distinct field of hydrology to study water, because the subject, in his view, had 'largely grown up in the families of sister sciences' and there was a 'tremendous pressure for hydrologic research created by recent activities in soil and water conservation'.[30] Horton grounded the discipline using the concept of the 'hydrological cycle'.[31] The hydrological cycle traced water from its evaporation as vapour over the ocean to its transport and storage over land, its ultimate precipitation (rain, sleet, snow), infiltration into the soil, run-off into streams, interception and evaporation from vegetation, and re-evaporation again from ocean and land. Secondary school students now study the hydrological cycle in textbooks, with illustrations depicting rain rising from the ocean, falling on land and mountains, then returning to the ocean.

Despite hydrology's vast theoretical domain – from underground aquafers to the skies and oceans – in practice Hortonian hydrology, as it was called subsequently, focused on what could be measured on or near the ground, 'the basic data', as Horton described it. Horton also wanted to ground hydrological research on fundamental principles from the well-established fields of physics and hydraulics to justify its credibility.[32] He knew the pitfalls of overstatement and controversy because he started his career as a district engineer in New York for the U.S. Geological Survey in the early 1900s, a period when debate over climatic botany reached a pitch in America. Horton participated in the debates by translating important Germanic forestry texts on transpiration and rainfall by Ernst Ebermayer, mentioned in Chapter Four, and Franz von Höhnel (1852–1920), but he kept his personal views close to his chest. In his translation of Ebermayer's work, Horton took a slight dig at 'enthusiastic conservation propagandists' but he did not reject their views entirely. In a non-committal manner, he suggested that 'there is no simple general rule as to the effect of deforestation which will apply to all climate, forests or drainage basins.'[33]

Horton's vision of hydrology tried to create sound policy by not allowing theoretical speculation to run ahead of empirical evidence and physical principles. He personally engaged with research on all aspects of the hydrological cycle, including atmospheric and oceanic influences, but he, like most other hydrologists, left the question of how water moved through the atmosphere primarily to meteorologists and climatologists. Commenting on the role of meteorology in hydrology in his 1931 article, Horton downplayed the atmosphere as a key preoccupation of water policy by suggesting that 'the economic applications of hydrology are derived almost wholly from the surface and underground waters.' This became a truism in water management: it is best to manage the water you have once it falls rather than try to change how much water you receive in precipitation. Hydrologists could measure a drought, but it was not necessarily their job to explain why the drought happened. By avoiding the question of where precipitation came from, hydrologists could focus on measuring how the hydrological cycle happened closer to the ground. Thus the majority of hydrological research in the twentieth century has focused on the Earth's surface, or merely a few metres above and below it.

Horton's work did, however, provide a glimmer of hope for forest–climate advocates. During the Second World War he published a paper on how the ocean potentially influenced hydrological dynamics on the land.[34] Intriguingly, he noted that the amount of chlorine found in surface water decreased the further inland one went from the ocean. For instance, California's or New York's water had more chlorine than South Dakota's. This led Horton to make two assumptions: first, chlorine facilitated the formation of condensation by providing stable nuclei on which to develop; second, the lack of chlorine in the inlands of the continent meant that much, if not the majority, of the precipitation from those regions came from regional evaporation and not from the ocean. As a result of this paper and other works, Horton was sometimes perceived as an advocate of rainfall recycling.[35]

Despite theoretical suggestions that a significant proportion of inland water was recycled, the field of hydrology undermined a key assumption on the forest-hydrological balance through a sustained assault on the idea that forests increased the availability of water. The U.S. government and university hydrologists played pivotal roles in policy planning from the 1930s onwards, putting a nail in the coffin as regards to orthodox beliefs that forests were necessarily good for water management. Government interest in water resources boomed during the 1930s because the droughts ravaging the country had led to declining water supplies.

In the United States, hydrology as a discipline gained authority and funding because of the rapid expansion of dams and water conservation initiatives led by the CCC and Army Corps of Engineers. Not only were some of North America's largest dams created in the 1930s, but so too were many of the United States' most important hydrological research experiments. The CCC built paired-catchment experiments at the Coweeta Experimental Forest in North Carolina in 1934. Paired-catchment experiments use two or more different streams to test how different landscapes influence streamflow. For instance, an experiment may take a non-forest grassland area and plant a forest on one stream while leaving another area unplanted to compare how the different treatments influence the amount of water in the streams. Coweeta and a number of other stations created at San Dimas, California, and Sierra Ancha, Arizona, sought to determine whether forests increased or decreased the water available within a catchment compared to non-forest vegetation types, such as grassland.

Despite much debate about the relationship between forests and streams, hydrologists in the 1930s had relatively little data to determine whether forests increased or decreased water availability in catchment areas. Governments and other scientists wanted to increase the amount of water in catchments to improve the availability of water for human use and, secondarily, to potentially increase precipitation. Two contrary views prevailed about the relationship between forests and water. The orthodox policy in the 1930s assumed that forests increased the overall availability of water in a catchment area. Forest cover was imagined to stop extreme flooding, improve retention in soil and stop the erosion of soil in high rainfall events. A smaller but vocal contingent of farmers, engineers and sceptical foresters argued the opposite: they believed that trees lowered available water in a catchment because trees used (transpired) water rather than letting it release into the soil or catchment. French engineers had been making this point since the nineteenth century, but it remained a minority view until it gained stronger traction in the 1940s.[36]

Early hydrological research proved inconclusive in determining the impact of forests on water. Research from Wagon Wheel Gap in Colorado, a high mountain research station that was in operation from 1910 to 1926, suggested that forests had no overall influence on streamflow. Forests neither increased nor decreased available water supply. The Wagon Wheel Gap study contradicted an earlier single catchment study in the White Mountains of New Hampshire that ran from 1910 to 1911. The hastily run, methodologically shoddy (even based on methods of the time) White Mountain study helped justify the Weeks Act. The Wagon Wheel study had the potential to overturn a century of orthodoxy. W. C. Lowdermilk (1888–1974), a prominent soil conservationist, called the study an 'attack upon the widely accepted belief that watershed vegetation must be kept intact for the most favorable influence upon streamflow and erosion and flood control'.[37] Nonetheless, the idea that forests protected water supply remained the orthodox position throughout the decade. In 1941 the CCC handbook stated simply, 'forest cover ... regulates the flow of streams.'[38]

Perhaps nowhere did the question of whether forests increased or decreased water availability matter more than in South Africa, an arid country deficient in both water and native forest cover. South Africa has historically had relatively few trees owing to its fire-dominated ecosystems of heath in the Cape, grasslands across much of the interior and savanna in the east and north. European colonialism, the discovery of diamonds and gold and mining-led industrialization all led to a rapid depletion of the country's limited indigenous forest cover in the nineteenth century that made wood one of the costliest imports into the region. Tree planting in South Africa was motivated by the demand for wood and a desire to change the climate. Foresters experimented with and planted trees with great gusto in the late nineteenth and early twentieth century, influenced by European and British Indian theories of forestry. The failure of plantations in the arid Karoo, an area inland of Cape Town, proved the first warning that trees could not make the deserts bloom. Over time, farmers complained that exotic trees, especially thirsty Australian eucalyptus, dried streams and did not increase rainfall. This debate simmered for decades among farmers, but it blew up in the 1930s, a time of drought and economic hardship. Since the mid-1920s, the South African government had embarked on a massive campaign of tree planting that created hundreds of thousands of hectares of eucalyptus, acacia and pine plantations on the hilly and mountainous headwaters of streams. Farmers downstream complained that their crops withered, and once abundant streams slowed to a trickle or, worse, ran dry.

Forestry became the subject of major political debate in South Africa. In 1935 South Africa hosted the Third Empire Forestry Conference, a meeting of foresters from around the British Empire who gathered to discuss science and policy findings. At the opening of the conference, Deneys Reitz (1882–1944), the minister of agriculture and forestry, told invitees that 'irate farmers' had accosted him because they believed exotic trees sucked up all the water in rivers.[39] The nation's prime minister, Jan Smuts (1870–1950), echoed the same views, warning, 'There is no doubt that a popular feeling is arising in South Africa that afforestation is causing the drying up of springs and water sources.'[40] Smuts and Reitz wanted foresters to investigate the question thoroughly and independently. A committee of foresters established by the conference enquired into the question and came to a somewhat different conclusion to the politicians and farmers: 'afforestation may have only slight bearing on the climatic conditions of a country,' they posited. Nonetheless, they continued, 'we cannot do otherwise than commend any and all efforts which are being made, or which can be made, to bring under forest cover a greater proportion of the land area of the Union.'[41] The committee suggested research should be done into the question because not all participants agreed with the report.

To answer the question, the South African Forestry Department established an experimental paired-catchment hydrological station at the Jonkershoek Valley in 1935, just a few minutes outside the university town of Stellenbosch and an hour's trip from Cape Town. The study sought to determine the longstanding question of whether trees increased or decreased overall water availability in the catchment area. The study area was in a closed mountain catchment that had the highest rainfall in the entire country. This was because the mountains collected oceanic-born winter rains. The site featured indigenous heath and a range of smaller streams flowing into the Eerste river. Christiaan Wicht (1908–1978), the director of the study, designed a paired-catchment experiment whereby he would plant pines near certain streams while leaving others in their native condition. The study would over time cut trees as well as burn indigenous heath to determine the impact that the harvesting and burning activities had on water supply. This way he could measure water in catchments before, during and after they had been forested, while also having the 'control' of indigenous vegetation.

What Wicht found at Jonkershoek turned forest orthodoxy on its head. In 1943 and again in 1949 Wicht concluded that planted trees in an area previously devoid of trees decreased the overall water availability in the catchment. Viewed from the point of transpiration, this argument made perfect sense. Trees use significantly more water than grass to grow and stand upright as a result of hydrologic tension; therefore, they also transpire larger amounts of water. Ground water storage, in fact, had relatively little to do with trees or soil but related more to soil composition, which determined the penetration of water through soil.[42]

A growing number of catchment-scale studies confirmed that forest cover decreased water availability in a catchment – that is, forests led to less water. The majority of studies came from the United States and Australia, but global examples, especially from South Africa, informed this conclusion.[43] A 1966 review of 39 different treatments – which included cutting and then planting forests along streams in a single catchment – suggested two clear findings. First, decreased forest cover 'increases water yield'.[44] Second, the 'establishment of forest cover on sparsely vegetated land decreased water yield'.[45]

Researchers also noted the complex interactions of specific sites – that is, that generalizations had varying outcomes depending on locale. The clear-cutting and burning of vegetation in an alpine Colorado pine forest only marginally increased streamflow, by 34 millimetres (1¼ in.), whereas in East Africa the loss of water-hungry bamboo forests increased waterflow by an amazing 457 millimetres (18 in.).[46] 'Demand-side' hydrological thinking put the final nail in the coffin of forest–climate beliefs that had lingered since the late nineteenth century. The idea that forests directly influenced rainfall stopped influencing forest policy in most countries around the First World War, but the belief that forests positively increased water in catchments – an essential part of local moisture models of precipitation – remained orthodoxy in places such as the United States, Europe and European colonies until the 1940s and '50s. As a result of paired-catchment studies, water managers turned their focus to regulating demand and the use of water rather than trying to influence the creation of new supplies of precipitation. Forests continued to be valued for wood and certain ecological benefits, an idea encouraged by incipient post-Second World War conservation biology, but they had little role to play in rainmaking.[47]

How Meteorologists Challenged the Evaporation–Rainfall Link

Meteorologists faced considerably different physical challenges when measuring water in the atmosphere compared with hydrologists, who only had to bend down or at most climb a ladder to measure it on the ground. Meteorologists had to understand atmospheric dynamics occurring kilometres above the Earth's surface. Instead of measuring large accumulations of water using rain gauges and weirs, meteorologists often found themselves chasing individual molecules that were invisible to the human eye.

Like the history of hydrology, there is an ancient lineage to thinking about moisture and the atmosphere.[48] Human fascination with the heavens may be as old as language and culture. It is well documented that the two oldest continuous cultures, that of the southern African Bushmen/San and Australian Indigenous Aboriginals, both have intricate procedures and knowledge of rainmaking. Rainmaking might be scoffed at by scientists today, but anthropologists since the late nineteenth century have recognized close connections between 'magic' and 'science': both suggest a naturalistic knowledge of the world, and both suggest that knowledge and techniques can be applied to manipulate the natural world.

Until the modern era, knowledge of the atmosphere came from on-the-ground human experiences. Once societies developed the technology required to fly, first with balloons and later with aeroplanes, this all changed. In the mid-nineteenth century, meteorologists started to gain an appreciation for the scale and problem of measuring atmospheric moisture following the advent of human-piloted gas balloons. Originally invented in France in the 1780s, by 1862 balloons were used by the British aeronauts James Glaisher (1809–1903) and Henry Coxwell (1819–1900) to voyage some 11,000 metres (36,000 ft or 7 mi.) into the air, the highest any human had flown. Wrapped in thick coats, the cold stung them bitterly. The pair nearly died from a lack of oxygen and gas pressure in the blood and brain at the apex of their flight.[49] Coxwell passed out, and Glaisher, who lost control of his hands while climbing on rigging, luckily pulled the string to release gas just before he succumbed. Glaisher gained public fame from this voyage (the third out of 28 during his career), but his reputation among meteorologists soared because of the data he recorded and its implications. He found that the air was not uniformly cold as one ascends: the air temperature varied. This variation suggested that different air masses overlapped.[50] The historian Peter Moore writes, 'This discovery would open up a line of enquiry that would occupy meteorologists for decades and would eventually lead to the delineation of the various stratum.'[51]

Unmanned balloons offered the potential to go even higher than piloted trips. Meteorologists sent sounding balloons higher into the atmosphere in the 1890s. Léon Teisserenc de Bort (1855–1913), a French meteorologist, collected data from 236 sounding balloon voyages he had sent from his station.[52] Whereas Glaisher's data did not show a simple linear relationship between altitude and height – he recorded variations as the temperature dropped during the ascent – Teisserenc de Bort's findings showed a layer of atmosphere that began between 8 to 12 kilometres (5–7½ mi.) above the Earth's surface that did display a linear relationship. In 1902 Teisserenc de Bort coined the terms 'stratosphere' to describe this highest atmospheric zone and 'troposphere' for the lower level. The word troposphere, in Greek, roughly translates to sphere of changes ('tropos' meaning change and 'sphere' referring to the Earth), while the stratosphere refers somewhat blandly to a sphere of layers. The names were chosen because the lower level showed greater variations in air mass and temperature, whereas the higher level showed continuous graduation.

The First World War not only saw the establishment of independent air forces – the British Royal Air Force, established on 1 April 1918, being the first – but acted as a catalyst for national and university meteorology programmes.[53] Weather forecasting gained renewed importance because storms or strong winds could hinder or help in battle. Militaries and universities rapidly rolled out the first specialized meteorological degrees near the end of the war. The U.S. Army established a Signal Corps school at Texas A&M University in 1918. The sustained support and interest continued into the 1920s. In 1928 the Massachusetts Institute of Technology (MIT) started the country's first graduate training programme in meteorology.

Aeroplanes and advances in physics paved the way for a better understanding of meteorological dynamics. The term 'modelling' is frequently used in media, and models are key parts of all climate and meteorological science today, but in the early twentieth century understanding of meteorology was largely descriptive and responsive. Significant shifts in barometric pressure remained one of the pre-eminent tools for predicting storms, but this tool gave relatively little warning of the advance of storms. Statistical meteorology laboured under an onslaught of facts and data. One prominent British meteorologist, Lewis Fry Richardson (1881–1953), during the First World War predicted it would require 64,000 people working full-time to predict the weather for the whole world.[54] Difficulties quantifying weather encouraged further research, especially into theoretical aspects of meteorology. Charles Marvin (1858–1943), chief of the Weather Bureau in the mid-1920s, encouraged American theoretical research on atmospheric events.[55] Physics was seen as one of the keys to unlocking atmospheric circulation and other secrets of the sky. If one understood how the atmosphere circulated, then predictions of rain, temperature and pressure could be made with greater accuracy.

One of the world's great centres of meteorological thinking from the 1920s to the 1940s was the Norwegian city of Bergen, located on the country's rugged fjord- and inlet-strewn west coast, which was home to large fishing fleets. In 1917 Vilhelm Bjerknes (1862–1951), a Norwegian physicist with backgrounds in fluid dynamics, mechanics and electromagnetism then working at the University of Leipzig in Germany, decided to come back to Norway to help establish a geophysical institute at the Bergen Museum. In 1917, in the midst of the First World War, Norway was a bleak landscape in which to start developing a new school of meteorology. The nation was suffering not only from the global conflict, but food shortages as a consequence of poor weather. The Norwegian government gave Bjerknes 100,000 kroner to support weather forecasts for farmers. Owing to wartime secrecy, which stopped the free flow of weather information via telegraph, Norway 'had to keep up [its] meteorological service without the accustomed help of weather telegrams from surrounding countries', especially Britain.[56] As a result, Bjerknes and his students had to investigate local weather phenomena without the aid of information from Iceland and Britain to predict serious storms.

Bjerknes helped establish weather monitoring stations throughout the country, with many located on tiny islands dotting the west coast. His son Jacob (1897–1975), himself a physicist and meteorologist, studied with great interest the lines of squalls, oceanic rain and windstorms that rolled with regular frequency into Bergen's port off the Arctic North Sea. Precipitation was a central, though indirect, focus of the Bergen school. Jacob paid particular attention to the formation of rainfall in cyclones as well as localized thundershowers over western Norway.[57] Predicting major storms could save the lives of thousands of fishermen living in small coastal towns who plied the waters daily to supply the nation with fish.

Jacob Bjerknes and his father discovered that when two different hot and cold air fronts converged, a storm would form in a cyclonic pattern. The mixing of air masses led to an ascending motion of air that caused precipitation to fall in a large frontal band, called the steering line, and a thin but intense rear flank, called the squall line, where heavy rain could fall. Later described as hot and cold fronts, polar front meteorology, as it became known, allowed meteorologists to better predict when the conditions for large North Sea storms were developing.

The Bergen school created a three-dimensional dynamic model of cyclonic activity that used fluid dynamics, physics and practical observations to explain the mixing of hot and cold air masses. This view gave further support to the existing theory that rain occurred when atmospheric conditions created ascending motions of air – that is, when moisture-laden air rose high enough in the atmosphere to form precipitation.[58] In 1923 Alexander McAdie (1863–1943), a Harvard meteorologist, summarized this view clearly in his popular book *Making the Weather*: 'It is the forced draft which counts.'[59]

The Bergen school created a model of meteorology that was followed by others in the 1920s and '30s. Its adherents applied the fundamental principles of hydrological and thermodynamics to explain meteorological processes . The findings of the Bergen school spread first throughout Scandinavia then later in Europe and North America. It succeeded where older descriptive synoptic charts had failed, as the discoveries made it possible to predict the early development of strong storm bands along the northern pole. Denmark adopted the Bergen method in 1921 after its national forecaster failed to predict a devastating October storm using old synoptic methods, which could not identify the convergence of high and low air pressure systems.[60] American meteorologists, too, adopted the Norwegian classification system for air masses. This system described air masses based on their 'source regions' (for example, arctic, tropical) and then divided them into 'maritime' and 'continental' conditions.[61]

Technological innovations in flight and radio across the decades improved the collection of data on meteorological dynamics. In the 1920s and '30s, the U.S. Weather Bureau initiated upper air research by establishing 26 stations across the country that produced daily analysis of the atmosphere up to 5 kilometres (3 mi.) in height using radio and aeroplanes.[62] American meteorological findings confirmed the views of the Bergen school that the movement of large air masses determined climatic and weather patterns. Meteorological researchers coalesced around a set of properties that could be measured in all air masses – humidity (the amount of moisture), temperature and pressure – which they used to theorize the dynamics of air masses, clouds and other meteorological phenomena.

New theories from the Bergen school and local data from the upper atmosphere measurements in North America helped meteorologists to further downplay local sources of moisture as the origins of precipitation. In 1937 Benjamin Holzman (1910–1975), a meteorologist working for the Climatic and Physiographic Division of the Department of Soil Conservation, then under the direction of the influential climatologist Charles Warren Thornthwaite (1899–1963), published an important technical bulletin for the U.S. Department of Agriculture that laid out what would become the established American meteorological position on the origins of continental precipitation over the United States. The paper reinforced Holzman's sterling reputation and became an important contribution to key moments in modern history, including helping to forecast the weather for both the D-Day invasion and the first nuclear test at Alamogordo.[63] Holzman is remembered today for his forecasting, but his paper 'Sources of Moisture for Precipitation in the United States' is recognized as the world's first analysis studying the origins of precipitation for a continent.[64]

Holzman referenced hydrological research in suggesting that less water returned to the sea than fell on the land as rain. He stated that 'from this it was inferred that continental evaporation was an important meteorologic process and constituted the principal source of moisture for continental precipitation.'[65] Thus 'the belief was readily accepted' that evaporation from the continent – trees, lakes, soil and rivers – produced a significant amount of America's moisture. Hydrological findings gave rise to the view that 'a mere increase in atmospheric moisture was sufficient for the production of local and other rainfall.'[66] These beliefs, he argued, were more readily accepted by 'earnest supporters . . . among hydrologists and silviculturists' than by meteorologists.[67] Holzman's 38-page thesis presented a comprehensive challenge to the idea that evaporation generated an appreciable amount of precipitation. He stated boldly, 'contrary to theories widely held, the major part of the moisture absorbed by these air masses is not precipitated on the land but is carried back to the ocean.'[68] Moreover, he argued that precipitation 'is derived not from land-evaporated moisture but chiefly from great maritime air bodies whose moisture is obtained by evaporation from oceanic areas.'[69] Holzman pointed to evidence that suggested that the amount of moisture found in the atmosphere had no direct influence on whether rain fell or how much would fall. For instance, measurements of moisture showed, surprisingly, that deserts often had sufficient moisture to produce precipitation, but the atmosphere lacked the proper conditions to create rain.

Research into upper-atmospheric dynamics also revealed where all the 'lost' terrestrial water went. Most of the air that was evaporated from continental sources was rapidly absorbed by large upper-air dry air masses that carried water vapour back to the ocean.[70] American meteorological researchers did not yet know about jet streams, a phenomenon discovered during bombing runs in Japan during the Second World War, but they understood that large dry air masses did exist and could move rapidly. American meteorologists classified air masses based on similar classification systems developed by the Bergen school.[71] Lest anyone think that a forest or lake could influence large-scale meteorological dynamics, Holzman pointed out that the atmosphere was governed by thermodynamic processes driven first and foremost by radiation from the Sun and, second, by radiation from continental land.[72] Air masses circulated globally between the cooler poles and the warmer equator. The interaction with large-scale air masses led some air fronts to absorb huge volumes of moisture from the oceans that, in moments of thermodynamic instability such as rapid upward movement of air combined with moisture, caused precipitation to fall. Holzman wrote that in 'the United States it may be said that the atmosphere practically always contains an ample supply of moisture for precipitation. The cause of the failure of rainfall to occur in a region, therefore, is the absence of conditions essential to the release of the atmospheric moisture.'[73]

Holzman challenged the dominant hydrologic view that much, if not the majority, of continental evaporated moisture fell again as precipitation over land. If the precipitation which fell on the ground came from evaporation primarily, then would droughts ever end without an infusion of new moisture?[74] Obviously the moisture needed to stop a drought came from somewhere, and if not the land, then it must have come from the ocean. Other evidence suggested that particles in air masses moved long distances, a fact that undercut the idea that water evaporation fell within the region where it came from. Holzman cited evidence from the Dust Bowl, where dark soil from the midwest was whipped up by winds and transported 3,200 kilometres (nearly 2,000 mi.) to New Hampshire and Vermont.[75] Even more striking was the long-range distribution of volcanic ash from the Indonesian volcano Krakatoa in 1883, which caused the entire Earth's temperature to fall because the ash blocked out the rays of the sun. These examples proved that particles – be they dust or water – could and did move long distances in relatively short periods of time.

Atmospheric readings of moisture further challenged the idea that the amount of water vapour in the air determined rain. Holzman pointed out that many advocates of rainfall recycling implied that the availability of moisture in the atmosphere determined the amount of precipitation that fell. Within this view, it was possible to increase the amount of overall water vapour via evaporation and transpiration. As logical as this seemed, Holzman pointed out that the average amount of water vapour in the air in San Diego, California, was the same and sometimes up to 50 per cent higher than in Eureka, California. Yet Eureka averaged just over 100 centimetres (40 in.) of annual rain compared to San Diego's 25 centimetres (10 in.). An even more extreme example could be found along the desert coastline of lower California, where on average 13 centimetres (5 in.) of rain fell per year (and then primarily in mountains) but the atmosphere, aided by oceanic air, had ample moisture to produce rain. Indeed, he argued that the entire United States had sufficient water vapour to produce rain on any given day. Holzman concluded, 'The failure of precipitation is not to be attributed to a deficiency in the supply of atmospheric moisture but rather to a lack of meteorological forces necessary to condense and precipitate the moisture.'[76]

Holzman's seminal paper attempted to put together a comprehensive theory for the origins of atmospheric precipitation, but due to limitations in data and technology his ideas had numerous shortcomings that even his advocates recognized.[77] Prior to the Second World War, upper atmospheric measurements remained rare and meteorologists could not measure differences in molecules with accuracy.[78] Even the basic communication of data on weather was limited – meteorologists communicated weather data to each other using code words rather than using large numbers because of the cost of sending information across the country.[79] As a result, Holzman relied on models rather than empirical evidence in many cases. He was unable to demonstrate that all continental precipitation originates from oceanic water vapour. Nor could he prove that all or even most of the evaporation from land was removed to the ocean before falling again on the land. Holzman lacked the ability to identify the origins of specific water molecules. Holzman's article was 'widely accepted' by American meteorologists due to an agreement on the dynamics of the atmosphere, but the lack of convincing proof meant that the idea that continental evaporation produced significant amounts of rain remained valid for some hydrologists and foresters.[80]

The 1930s also witnessed another important conceptual development that helped to explain the formation of precipitation. If the atmosphere had sufficient water vapour, what then explained why precipitation formed and fell? Also, what causes snow, rain and sleet? To explain this, one European meteorologist looked to a local phenomenon – fog over icy roads. Tor Bergeron (1891–1977), an influential Swedish meteorologist who represented the Bergen school, gained inspiration for his theory of precipitation by observing days when fog hung over the roads around the Voksenkollen health resort.[81] In winter, he noticed that the roads were foggy if the temperature was above freezing but when the air temperature reached freezing point the roads became clear. This dynamic inspired him to build on the theory of Alfred Wegener (1880–1930), a German Arctic researcher most famous for his theory of continental drift, which suggested that precipitation required both supercooled water vapour and ice. Supercooling occurs when the temperature of water vapour goes below freezing without turning into a frozen solid. The tiny, supercooled water particles lack a nucleus core on which to form crystals. Supercooled droplets are tiny – hundreds of thousands or a million of them are required to form a single snowflake – and they will not fall on their own. To form and fall as precipitation, these supercooled water particles need to attach to nuclei.

Bergeron's theory suggested that when the supercooled air met nuclei, the moisture attached to nuclei and the added mass of nuclei caused the moisture to fall to the ground. If the air was freezing, this precipitation fell as snow. If the air temperature was slightly above freezing it could fall as sleet, and if the air temperature was warmer than freezing it would fall as rain. Experimental research in

Germany by Walter Findeisen (1909–1945) later in the decade advanced Bergeron's theory into a coherent theoretical framework that was accepted by the meteorological community.[82] The Bergeron-Findeisen theory, as it became known, remains a primary mechanism for explaining how precipitation forms. In 1949 Bergeron further developed his idea by suggesting that precipitation could be 'seeded' by layering ice crystals, which would provide nuclei for the super-cooled water in the atmosphere to bind to and form precipitation.[83] This idea provided inspiration for efforts to modify rainfall after the Second World War.

The Cold War era led to a blossoming of support for meteorology in the United States, the world's leader in the field in terms of funding and research. University courses in meteorology flourished during and after the Second World War at centres such as MIT, the University of Chicago, New York University and the University of California, Los Angeles, in order to produce a new cohort of graduates to take on the numerous government, university and private-sector jobs.

Because of post-war financial support and advances in technology, meteorologists from around the world received a range of opportunities hitherto only imagined – from studying how nuclear bombs destroyed a layer of the upper troposphere to cloud-seeding experiments that dumped tons of chemicals onto clouds. For obvious reasons, meteorologists working for superpowers such as the United States and later the Soviet Union, Britain and France had access to cutting-edge technology and classified military intelligence. The development of nuclear weapons during the Second World War created a new 'environmental catastrophism' that fuelled research into theoretical and practical problems of the atmosphere.[84]

The question of how precipitation formed was central to major national and international projects in the 1940s to the 1960s. At its most sinister, military officials concocted ideas such as manufacturing droughts to cripple the energy and food supplies of enemies. During the Vietnam War, the American government's Operation Popeye tried to enhance the rainfall in Vietnam and Laos to create chaos among communist supply chains.[85] At its most capitalistic, private corporations, including General Electric, tried to profit from learning how to control the weather using dry ice and silver iodine to turn rain on and off with the simple flip of a switch. But in reality, the majority of meteorologically aligned researchers – who came from diverse fields such as hydrology, climatology, meteorology, engineering and physics – really just wanted to learn about how the skies worked.

A few years after the Second World War ended, the University of Chicago-trained meteorologist George Benton (1917–1999) and two colleagues drew on the growing upper-atmospheric data to estimate the flux of moisture over the Mississippi basin, an area covering approximately 2.6 million square kilometres (1 million sq. mi.) from Minnesota to the Gulf of Mexico. Their influential study made two key findings that further reinforced Holzman's view. First, they observed that less than 20 per cent of the moisture in the maritime air coming from the Gulf of Mexico fell as precipitation. A staggering 80 per cent of ocean-derived moisture remained in the sky and eventually returned to the ocean without falling as precipitation at all. Second, and most importantly for the origins-of-rain question, their analysis suggested that local evaporation from the land contributed only 14 per cent or less of the overall precipitation in the basin. The authors concluded that nearly 85 per cent of all the rain in the huge basin came from moisture from the Gulf of Mexico and the Atlantic Ocean. Benton and his colleagues concluded: 'Increasing or decreasing precipitation even slightly by local regulation of land use is therefore out of the question.'[86] A different study in West Virginia by Benton concluded similarly that over 80 per cent of precipitation came from maritime air and less than 10 per cent could be attributed to evaporation.[87]

Benton's study became a cornerstone of the meteorological view held by American, European and Soviet scientists that evaporation played only a small role in the formation of precipitation. Earth's rain and snow came from maritime air. Benton's and Holzman's findings raised an intriguing question. If most of the moisture in the air never fell as precipitation, could humans find a way to induce the moisture to fall as rain? The realization that the atmosphere had, at almost all times, sufficient moisture to produce rain encouraged a small group of scientists and an even larger number of American politicians and supporters to imagine that, with a few technological fixes, rain could be turned on and off like a light.

Early successes in artificial rainmaking bolstered this view. Only a few years after the war ended, the Nobel Prize-winning scientist Irving Langmuir (1881–1957) and his colleagues Bernard Vonnegut (1914–1997) and Vincent Schaefer (1906–1993) proved that it was possible to 'seed' precipitation-bearing clouds. In his lab at General Electric Laboratories in upstate New York, Schaefer, an expert machinist who spent the Second World War devising studies for creating smoke screens and de-icing aeroplanes, found that rapid changes to temperature – in this case caused by putting dry ice into a box – generated millions of tiny ice nuclei, the structure required to turn supercooled air into precipitation. Schaefer and Langmuir capitalized on this finding because during the Second World War they had developed experiments and equations to predict the development of supercooled droplets that formed ice to protect aeroplanes from crashing in winter.

Drawing on the Bergeron-Findeisen theory of supercooled water and ice crystals, Schaefer, Langmuir and Vonnegut proposed a method for seeding clouds on a much larger scale. If the atmosphere did have self-sufficient moisture to produce precipitation, as Holzman and other meteorologists suggested, then all that would be needed to generate rain or snow was a catalyst to induce the right meteorological conditions. Their successful cloud-seeding experiments, first with dry ice and later with silver iodine smoke, captured the attention of meteorologists and militaries throughout the world. In 1951 Vannevar Bush (1890–1974), leader of the Manhattan Project and instigator of the prestigious American National Science Foundation, stated, 'we are on the threshold of an exceedingly important matter, for man has begun for the first time to affect the weather in which he lives, and no man can tell where such a move will finally end.'[88]

The first efforts by researchers at cloud-seeding used aeroplanes to drop dry ice directly onto clouds with sufficient supercooled moisture. The United States, the Soviet Union and Australia all quickly developed research programmes to seed clouds. Taffy Bowen (1911–1991), a Welsh physicist who worked at Australia's Commonwealth Scientific Industrial and Research Organisation's (CSIRO) Division of Cloud Physics brought cloud-seedings to Australia in 1947. In early February 1947 the Royal Australian Air Force dropped dry ice from aircraft onto a cumulus cloud formation. The cloud expanded and produced 12 millimetres (½ in.) of rain, whereas the other neighbouring clouds did not. This experiment indicated to researchers elsewhere that seeding could generate appreciable amounts of precipitation. The early use of dry ice proved too costly and bulky, however, so a variety of other chemicals were studied for use as nuclei.

Vonnegut, who worked with Langmuir at General Electric Lab, identified a mixture of silver and iodine as a better chemical structure to catalyse precipitation. Researchers tried using burners on the ground, but this often failed because the smoke was unable to reach the necessary altitudes and the silver iodine nuclei were damaged by sunlight. Eventually researchers burned silver iodine from the wings of aeroplanes. Initial results on cloud-seeding provided grounds for optimism, but the outcome of trials in the 1950s and '60s suggested that it had limited impact or possibly detrimental consequences on rainfall.[89] The areas where seeding seemed to work best were in high snowfall environments, such as the headwaters of rivers in Australia and the American West. Mainstream meteorologists benefited greatly from government support for cloud-seeding but the research was always hampered by atmospheric variability and the sheer magnitudes involved.

A government report noted in 1977 one fundamental problem: 'the atmosphere is an uncontrolled laboratory.'[90] It proved extremely difficult to determine whether rainfall was 'natural' or influenced by human activity.[91] After all, there is no way to measure whether rain would have fallen had cloud-seeding not happened. Weather does not happen in a controlled laboratory environment but across vast scales that cannot be replicated or measured for independent variables. Supporters of cloud-seeding often came from non-meteorological backgrounds or they had somewhat marginal reputations in the field. Two of the strongest American advocates were not trained meteorologists at all: Langmuir was a chemist and Schaefer a mechanist.

Langmuir and Schaefer, the leading advocates of weather modification, pushed hard to sustain government support and interest in cloud-seeding as a method of controlling weather despite scepticism from the start. A significant percentage – perhaps the vast majority – of meteorologists never saw weather modification as a viable practice except in limited circumstances. Roscoe Braham (1921–2017), a leading expert on thunderstorms and induced precipitation who worked in the field in the late 1940s to 1960s, later reflected, 'most, I think, were probably skeptical that . . . you could cause wide ranging and highly significant changes' using information inferred from such practices.[92] Many of Langmuir's colleagues, as well as historians, have judged these efforts critically because they opened the door for a number of opportunist meteorologists who started private businesses 'selling' rain. During drought in the early 1960s, commercial rainmakers touted experiments by Langmuir's team as well as results from Australia to promote their services to farmers.

Despite great fanfare, efforts to control the weather did not practically advance far beyond original work done in the 1950s. In 1969 Schaefer lamented to fellow advocates in the Weather Modification Association newsletter, 'As I watch the development of weather modification activities in the United States, I become increasingly dismayed at the lowering efficiency and effectiveness.'[93] Schaefer believed the lack of success in weather modification was caused by urban scientists who sat around computers rather than spending time in the air and working with their hands as he had done. Schaefer, a trained mechanist, derided the claim that meteorological dynamics were too complex to easily understand and control. He also complained that the public was unwilling to support efforts to control weather due to the possible problems, such as drought or extreme flooding. He criticized the 'attitude . . . that there are vested rights to the atmosphere and that natural phenomena such as "weather" should not be tampered with under any circumstances.'[94] This view reflected various court cases brought by plaintiffs whose property they believed had been damaged by weather caused by cloud seeding, and a powerful segment of public opinion influenced by a range of farmers, environmentalists and concerned citizens who worried about changing the natural cycles of weather. These concerns continued into the 1970s. A major flood in 1972 in Rapid City, South Dakota, which potentially occurred as a result of cloud-seeding experiments supported by the Department of the Interior's Project Skywater, killed more than two hundred people.[95] This only added to the public scepticism about government efforts to control the weather.

Instead of rejecting complexity, many leading meteorologists embraced it. James McDonald (1920–1971) published an important chapter on problems with the evaluation of weather modification field tests. The same year that Schaefer lamented the state of weather modification, McDonald noted the near impossibility of measuring cloud-seeding: 'our atmosphere is a physical system characterized by many degrees of freedom and exhibiting enormous variability. And it defies attempts at execution of the type of "controlled experiments" whose effectiveness we so admire in many of the physical and engineering sciences.'[96] McDonald was just one of several strong meteorological critics who questioned congressional funding that 'seem[ed] to constitute almost unreasonable munificence'.[97]

The hope that cloud-seeding would change nature fits within a larger history of rainmaking. Kristine Harper's concept of the 'entrepreneurial scientist' is useful for explaining the continued interest in weather modification by the U.S. government and a handful of scientists. Harper argues that 'politicians, with the aid of entrepreneurial scientists, were (and are) attempting to use it for their own political ends . . . But those entrepreneurial scientists and their scientific and technological expertise [are] not playing on the same level as state officials calling the shots.'[98] This is to say, military officials and enthusiastic politicians used science, and willing scientists, to investigate questions that mainstream scientists recognized were unlikely to pay dividends. James Fleming, the world authority on the history of American meteorology, pours cold water on scientists who bought into these 'fantasies'.[99]

Meteorologists could learn the secrets of the sky, but creating rain remained one step too far. Like other dreams of contending with and overpowering nature – such as rain following the plough and efforts to create forests to make the deserts bloom – humans found themselves thwarted in their attempts. Advances in technology, such as the nuclear bomb, allowed humans to destroy nature on an unprecedented scale, but it could not increase rainfall by even a small percentage.

Lake Fallacy

It is fitting to close the chapter with a closer focus on McDonald, a meteorologist who fought against cloud-seeding boosters such as Schaefer while he sought to put down once and for all the lingering popular view that evaporation produced rain. In his tongue-in-cheek 1962 article 'The Evaporation-Precipitation Fallacy', he chided the popular belief that lakes could create rain for the 'gross misconception concerning the hydrological magnitudes involved'.[100] Ironically, the Institute that McDonald worked for had been partly funded by wealthy ranchers in Arizona who wanted to increase rainfall through cloud-seeding. Despite meteorological consensus that most precipitation came primarily from oceanic sources, the belief that large inland lakes or tree planting could induce rain remained popular in the arid American southwest and even in the highest circles of the U.S. government.

McDonald had grown tired of reading opinions in local newspapers calling for the creation of a large inland lake in southern Arizona and he read with horror proposals to flood the Qattara depression in Egypt.[101] This idea began in the 1920s and remained an attractive one for the Egyptian government and its Cold War suitors, the Americans and the Soviets. As late as 1964, the reputable British magazine *The Economist* reported rosily that, if flooded, 'the new lake will have the effect of changing the climate of the whole area; the prevailing west winds should blow rain clouds against the northern ridges of the depression.'[102] Some naive enthusiasts imagined that the lake would secure Egypt's peace, thus ensuring stability across the Middle East.

McDonald invented a satirical example, Lake Fallacy, to illuminate the sheer impossibility – in financial, technical and scientific terms – of creating more rain in dry environments using artificial lakes. He pointed out that meteorological research showed that there was at least a ten-day turnover time from when a water molecule enters the atmosphere to its fall to earth as precipitation. Even if Lake Fallacy did increase atmospheric moisture that fell as precipitation, it would water Texas or Florida rather than Arizona. Moreover, a lack of rain in a locale was not necessarily explainable by a lack of moisture. Even in the worst drought, research showed that the atmosphere contained enough moisture to produce precipitation. Instead of lacking moisture, the paucity of rainfall reflected a lack of 'dynamic processes capable of producing ascending motions'.[103] He referred to the fact that rain falls on the windward side of mountains because upward air movements trigger precipitation. Increasing southern Arizona's rain by 10 per cent would require creating a lake as big as Lake Erie, one of America's Great Lakes, which borders Michigan, Ohio, Pennsylvania, New York and Ontario in Canada – an obviously impossible proposition. Cheekily, McDonald wrote:

> if any reader remains unconvinced, and obstinately sets out to excavate a large depression to be flooded for purposes of increasing local precipitation, he should take great care to pile all of his excavated dirt entirely in one area and not spread it around thinly. If he does so, and if he digs and piles long enough he can be sure that about the time his debris stands some 10,000 ft high, his rainmaking enterprises will actually begin to pay off.[104]

McDonald's views nicely summarized the attitudes of meteorologists in the 1960s: local evaporation had very little influence on local rainfall.

By the 1970s, the orthodox meteorological position that local evaporation played only a small role in precipitation became the established one in forestry and hydrology, the two disciplines that in the past had vigorously argued the opposite. The British hydrologist Charles Pereira (1913–2004), one of the world's top authorities on hydrology in the 1960s and '70s, concluded in his authoritative work *Land Use and Water Resources* that studies from around the world 'all agree that over large land masses only about 10% [of moisture for precipitation] is derived from vegetation and freshwater surfaces.'[105] Forestry researchers came to a similar consensus. A comprehensive review of forestry and hydrology in 1970 determined that the maximum potential influence forests had on rainfall was no more than 5 per cent.[106] For a time, the evaporation–precipitation connection had been put to rest. Meteorologists, hydrologists and foresters had finally formed a consensus on an issue that had lasted for hundreds of years. But for how long would that consensus last?

7

THE REVIVAL OF THE FOREST–RAINFALL CONNECTION

Open any contemporary life sciences textbook and you are likely to find a depiction of the water cycle. Pictures usually represent the cycle like this: first, moisture evaporates from the oceans; second, the moisture moves over land; third, the moisture falls on mountains and hills and flows down through rivers and lakes before, fourth, it finally returns to the ocean. But this is not the whole story, and a growing number of hydrologists and climate researchers have expressed concern about these common representations of the water cycle. In their view, these visualizations create false impressions that humans – and forests – have little to no influence over the water cycle. A 2019 study of water cycle visual descriptions found that 95 per cent of them feature a single catchment rather than looking at continental or regional scales.[1] The focus on a single catchment, the authors of the study argued, obscures the process of 'continental moisture recycling', a process driven by transpiration from plants and evaporation from the land and water. The paper represents the views of water and climate scientists who argue that forests recycle rain, often in significant quantities. The researchers seek to challenge the orthodox mid-twentieth-century hydrological and meteorological views on precipitation, which suggest that vegetation plays only a minor role in precipitation cycles. Once again, forests are being viewed by a now established and growing group of scientists as fundamental generators of rain and guardians of climate.

This final chapter examines the revival of forest-precipitation ideas beginning in the 1970s. New insights from the fields of hydrology, meteorology and climate modelling led a growing number of scientific researchers to start seeing vegetation as an important influence on a region's rainfall. Based on developments after the Second World War, the use of mass spectrometers to measure the mass of molecules gave hydrologists, meteorologists and oceanographers better insights into how water molecules moved through the water cycle. From 1961 scientists began creating a global catalogue of the composition of 'meteoric waters', a process which allowed scientists to more accurately trace the weight and movement of different water molecules throughout the hydrological cycle. Research on meteoric water in the Amazon basin in the mid- to late 1970s provided compelling evidence of large-scale moisture recycling in the region.

A separate process of climate modelling indicated that forests, such as the Amazon forest, played critical roles in mitigating human-induced climate change. By the late 1970s, advances in climate modelling began to produce more dynamic and accurate models that predicted how deforestation might impact on regional and global conditions, such as precipitation, humidity, temperature and albedo (the reflectivity of surfaces). This was the culmination of a process starting at the end of the Second World War that saw the emergence of computing models for understanding the complex dynamics of global and regional climate change.[2] By the end of the decade, many climate scenarios modelled by researchers indicated that large forests play a significant role in generating precipitation in tropical rainforests, especially the Amazon. At the same time, scientists that were focused on the world's growing carbon emissions began to worry that deforestation would further increase carbon dioxide in the atmosphere. A steady drumbeat of forest climate research focusing on carbon, precipitation recycling and albedo suggested that deforestation could have devastating impacts on the climate.

In the past ten years, a growing number of researchers have become increasingly vocal about the need to revise our thinking about forests and water at regional and global levels. In 2017 David Ellison, a leading researcher on the subject, and an international team of water and forest researchers made a 'call to action that targets a reversal of paradigms . . . to one that treats the hydrologic and climate-cooling effects of trees and forests as the first order of priority'.[3] They argue that the water-recycling properties of forests are underemphasized in current policies, which focus primarily on carbon rather than the wider context of climate. This statement coalesces a larger body of research that sees forests as 'biotic pumps', 'generators' or 'recyclers' that increase the overall water balance in and across regions.[4] These scholars boldly call for a 'timely resolution' to the two-hundred-year-old debate about whether forests are net users or suppliers of water in the hydrological cycle.[5] Ellison and others want to revise hydrological policies to allow for a significant expansion of forest cover, policies which would require both the protection of existing forests and the planting of new ones.

This position challenges a hydrological orthodoxy, which they describe as 'demand-side thinking', that sees water as a finite ecological service – a zero-sum game – that must supply competing users, ranging from people using water for drinking to farmers using water to irrigate their farms or workers in the mining industry needing water for their operations. If any user 'takes' water from an existing source, then the available water is diminished for all other users. The demand-side model does not see trees or any other land-use types as generators of rain – all users take water without putting any back into the system.[6] The demand-side model sees forests as particularly heavy users of water rather than as generators. Demand-side policies are based on long-term experimental and observation research projects in paired catchments that ran for decades and were designed specifically to inform policies for managing water and forests.[7] The projects did not focus on the question of where the moisture that caused local precipitation originated from nor did demand-side scholars seek to answer what happened to the evaporated and transpired moisture when it returned to the atmosphere.

As it stands, the demand-side model remains the dominant water-management framework in many regions, but it is being challenged and replaced by supply-side ideas. Supply-side scholars critique the demand-side perspective for focusing too narrowly on single catchments without properly accounting for the regional and global dynamics that influence the generation of precipitation.[8] Major reports by meteorological and climatic divisions of the United Nations, European Union and national governments have requested a rethinking of how forests provide the critical 'green infrastructure' that underpins agriculture and climatic stability. They call for the protection of existing forests and the expansion of new forests, which can increase precipitation, modify temperature extremes and provide forest products for the economy.

The implications could be significant. A major change in the scientific logic underpinning these policies would require a transformation in how industries, governments and the public understand forests and water. Hydrological policies based on demand-side thinking encouraged the view that trees are net users of water, whereas supply-side thinking sees trees as contributors to precipitation. If the reversal of opinion occurs on a global scale, it would encourage more reforestation and afforestation projects. However, this raises the question of how new and existing forests would be managed. Will they be harvested for economic purposes, or will the forests be protected as habitats for biodiversity and as carbon sinks, a site where carbon is taken out of the atmosphere and sequestered in wood and soil?[9] Or would they be used for a third path, that of semi-native forests which could not only be exploited for resources but used to expand biodiversity?[10]

From a historical point of view, a few aspects of these debates require further discussion. In scholarly investigations of atmospheric recycling, scientific researchers and policy experts show little awareness of past scientific ideas about forests and rain. Nor do studies promoting supply-side thinking examine the policies and consequences that flowed from older ideas of the link between vegetation and precipitation. The history of ideas of the relationship between forests and climate change is incredibly relevant to designing contemporary forest policies using supply-side thinking, not least because solutions put forth by previous generations – namely forest preservation and tree planting – are being proposed again today. That being said, there are obvious differences between past attempts at such policies and present-day endeavours, such as a greater awareness of how to sustainably plant trees and an appreciation for Indigenous rights and ecological management. But there are parallels that warrant closer scrutiny.

On Models and Tropical Rainforests

To understand the revival of forest-precipitation links, we need to explore the role that tropical rainforests have played in hydrological research and popular environmentalism. Scientific documentaries, popular articles and television news stories frequently portray tropical rainforests such as the Amazon as the 'lungs of the Earth', as a veritable Noah's Ark of biodiversity and as home to large numbers of Indigenous people living pre-modern lifestyles. Concerns about rainforests have played a significant role in shaping global environmentalism since the 1980s: first, through alarms about the loss of biodiversity, and second, because of fears that deforestation released large amounts of carbon into the atmosphere.[11] Research on tropical rainforests has played an equally significant, though less publicly recognized, role in shaping contemporary theories about how forests can influence local precipitation cycles.

The term 'rainforest' – which people frequently conflate with 'tropical rainforest' – has come into popular use since the 1980s. Indeed, Merriam-Webster's dictionary defines a 'rainforest' as a 'tropical woodland with an annual rainfall of at least 100 inches'.[12] Yet this definition is inaccurate because it leaves out other rainforests, such as the vast temperate rainforests of Douglas fir that form along coastlines of the Pacific Northwest of North America. It is important to distinguish between temperate and tropical rainforests for several reasons. First, temperate rainforests are much smaller than their tropical counterparts: forests within the tropics of Cancer and Capricorn – located respectively at 23 degrees north and south of the equator – that receive 200 centimetres (80 in.) of rain on average per year cover around 10 per cent of the Earth's land surface. Second, the large size of tropical forests means that they have the potential to produce significant climatic effects on a region. This is why researchers began to focus on tropical rainforests in the 1980s to understand how deforestation might impact on regional climatic systems. As modelling became more global in scope, the role that rainforests played in the global climate cycle grew in importance.

There is no simple tropical environment. Tropical environments range from arid deserts to high mountains, lush equatorial forests, arid savanna and vast grasslands. Despite this, popular conceptions of rainforests have often conflated the tropics into a singular entity. David Arnold's historical research on the tropics shows that the idea of the tropics emerged from European colonial expansion in the mid-1700s.[13] Most depictions of the tropics were negative, with the vast region being viewed as having unhealthy climates for European peoples. Tropical exoticism, the perception that the tropics were an unspoilt Eden of primitive forests, fit within a Rousseauian desire to imagine non-European lands and peoples as 'pre-historic'. Today, the idea of tropics as an unhealthy place has receded but we are still left with the popular view – promoted heavily by tourism marketers – that the tropics are exotic and untouched.

The idea that tropical forests increase rainfall goes back as far as the late 1400s when Christopher Columbus argued that dense tropical foliage created the humidity and the daily tropical showers he witnessed in the Caribbean. This idea maintained popular scientific appeal for the next five centuries.[14] Even critics of climatic botany who did not believe forests increased rainfall often created an exception for tropical forests. In *Man and Nature* (1864), George Perkins Marsh noted the views of then leading French foresters who agreed that rainforests created unique climatic conditions. The French forester Jules Clavé (b. 1828) argued that subtropical and tropical forests created more rainfall than in temperate regions: 'the action of forests on rain, a consequence of that which they exercise on temperature, is difficult to estimate in our climate, but very pronounced in hot countries.'[15] Antoine Henri Becquerel (1852–1908), a prominent French forest scientist, also considered it 'certain' that tropical rainforests induced rain but he could only venture that it was 'highly probable' that temperate forests stimulated a similar climatic impact.[16]

Scientists continued to carve out exceptions for tropical areas even after meteorological criticism dampened enthusiasm about the link between forests and rain in the early twentieth century. Julius von Hann, himself a sceptic on the connection between forests and climate, nonetheless believed that tropical forests increased rainfall. In his 1903 version of the *Handbook of Climatology*, Hann wrote, 'it may be concluded, with a great deal of certainty, that, so far at least as the tropics are concerned, forests may actually increase the amount of rainfall.' Hann noted that rainforests contradicted his normal view that, in 'general, the rainfall is to be looked upon as the cause and the condition of the cover of vegetation as the effect'.[17]

Throughout the mid-twentieth century, climatological thinking saw climate as determining vegetation (rather than the other way around), although researchers continued to point to exceptions in tropical forests. W. G. Kendrew's book *The Climates of the Continents*, originally published in 1922 and the foremost English-language textbook on climatology at the time, argued that rainfall was caused by geographic and climatic features, although Kendrew made an exception to this overall rule with the case of equatorial forests. He wrote: 'The heavy rainfall of the Niger delta and similar low tropical shores results in part from the increased [atmospheric] turbulence over the forests.'[18] Despite this exception, Kendrew's book reflected the increased dominance of the idea that large-scale atmospheric forces were driven by radiation from the Sun and moisture from the ocean.

The idea that tropical forests had distinct climatic traits continued into the second half of the twentieth century, a period that saw a global reordering of meteorological and hydrological knowledge of tropical climates. Decolonization and development in tropical countries created the conditions for the emergence of home-grown researchers who challenged dominant North American hydrological and meteorological orthodoxy. Before decolonization, European colonial scientists dominated thinking about water in the tropics. Most tropical researchers lived and worked for their entire lives in the temperate northern hemisphere and relied on data collected for them by researchers in tropical regions. There were a few exceptions, such as the influential hydrologist Charles Pereira, who spent two decades working for the colonial service in tropical east and southern Africa before returning to Britain in 1969.

The idea that forests create rain remained popular in tropical regions well into the mid-twentieth century, even when it declined in popularity among scholars in temperature regions. For example, the forest–climate connection served as a key strategic axiom for the small island city-state of Singapore. The newly independent Singaporean government embarked on a large-scale tree-planting campaign in the early 1960s to increase rainfall, as Singapore's lack of local water supplies threatened the country's viability. During the Second World War, the British Army surrendered Singapore because the island lost access to its water supply, which relied heavily on reservoirs connected via an aqueduct to mainland Malaya. In 1963 the prime minister for Singapore, Lee Kuan Yew (1923–2015), called for trees to be planted every year to increase the rainfall. Yew believed that politically Singapore had to separate from Malaysia, but such a decision would diminish the city's water supply. Chor Yeok Eng, secretary for the Ministry of National Development, called directly on Singaporeans to plant trees to 'induce more rainfall'. He argued that 'It has been scientifically proven that where trees are plentiful increased rains will result.'[19] Inche Othman Wok, minister for social affairs, emphasized, 'Quite apart from giving pleasant shade and enhancing the scenic beauty of the locality, trees also play a very important part in preventing soil erosion and, above all, in maintaining the normal rainfall.'[20] Government officials worried that the destruction of forests, exacerbated by a building boom, had worsened the drought that occurred in 1963. In 1971 Yew established Tree Planting Day on 7 November to further promote his tree-planting agenda.[21] Few people today know that Singapore's now famous garden city status was started partly to induce precipitation for the city.

Singaporean officials faced a problem experienced throughout the tropics in the post-war period. Rapid development led to extensive destruction of indigenous forest, especially wet rainforest. In the post-Second World War period, the wet tropics saw the highest rate of deforestation in the world. These vast rainforests – and the deforestation that ate away at their integrity – became global laboratories in the 1970s and '80s. Much is known about the expansion of research on biodiversity and extinction that occurred at this time, but almost nothing has been written by historians on how research on rainforests revolutionized hydrological thinking.

Brazilian researchers in the late 1970s and early 1980s came up with startling findings that raised profound questions about the environmental impact of tropical deforestation. Brazil's resource-intensive economy grew at an average rate of 7 per cent per annum during the 1960s and '70s.[22] This rapid growth partly reflected the conversion of forest to agricultural land and cities.

In the 1970s, Brazil's government made significant investments in applied scientific and technological research as part of widespread import substitution and development programmes that sought to increase economic growth. Meteorology received increased funding, in part because it supported the military government's effort to understand the climatic and meteorological conditions of the Amazon, which government officials wanted for development purposes rather than to preserve the vast rainforest.[23] Brazil had invested in meteorological research since the 1890s, when the country first established a meteorological service by installing a meteorological station in the middle of the Amazon, in the city of Manaus, in 1893.[24] By the 1970s, meteorologists began having access to new data on upper atmospheric measurements, in Brazil and in other tropical areas of the world. These aeronautical readings reflected the growing government presence over the airspace of the Amazon. Brazil's aerospace industry took off in the 1970s after the Brazilian Junta government established the Empresa Brasileira de Aeronáutica (Embraer) in 1969. Aeroplanes helped the Brazilian government to create accurate maps of Amazonia and to fly people into remote parts of the tropical region.

Starting in the mid-1970s, a cohort of young Brazilian researchers published important findings suggesting that the Amazon forest generated much of its local precipitation. This challenged not only meteorological and hydrological orthodoxy, but the Brazilian government's desire to open up the Amazonian basin as a site of economic development and population resettlement. Prior to the 1960s and '70s, the Amazon had remained relatively unscathed from the deforestation that had destroyed the vast eastern seaboard forests of Brazil starting in the 1500s.[25] In the 1970s, Brazil's government began to build motorways and other roads connecting the Amazon to the eastern and central parts of the country as part of larger plans to settle people and create farms in the vast rainforest.[26] The government hoped to relocate people from the populous but drought-prone northeast into an environment that had more rain and potential for increased agricultural output. This early settlement plan largely failed, but it marked one of many state attempts, continuing to this day, to move people into the forested interior.

Brazilian researchers benefited from training and collaboration with North American researchers, but they acted as lead researchers for much of the most important work. The American-trained Brazilian meteorologist Luiz Carlos Molion published a PhD dissertation in 1975 while studying at the University of Wisconsin-Madison that analysed the Amazon's water and energy balance using data on wind and humidity. His thesis drew from extensive meteorological and climatological data on the circulation of the atmosphere in the tropics.[27] Molion's study provided strong theoretical foundations for explaining that the Amazon forest is a large system that generates precipitation from evaporation and transpiration. His work suggested that up to 50 per cent of the rainfall in the Amazon basin could be produced by rainfall recycling.[28] The argument raised provocative possibilities that the Amazon forest did not fit the model established by researchers in the temperate northern hemisphere.

Two years later, in 1977, a Brazilian study of vapour fluxes – based on similar methods used by Benton and other American researchers in the 1950s – challenged the then orthodox American meteorological view that evaporation and transpiration had little impact on the overall amount of precipitation that fell. The study examined data on upper atmospheric moisture stretching from Manaus, in the

Amazon, to Belém, an eastern port city on the Pará river.[29] The research team, led by José Marques, drew on precipitation and air temperature records from 1972 compiled by the Ministério de Aeronáutica, the Ministério da Agricultura and the Instituto Nacional de Pesquisas da Amazônia (INPA). Their conclusions validated Molion's theoretical modelling. The team found that 52 per cent of rain came from the Atlantic and 48 per cent was recycled precipitation: almost a perfect 50:50 ratio. The implications of this study suggested that deforestation would lead to declines in rainfall.

Alarms that Amazonian deforestation could have a devastating climatic impact on the region started to appear in leading scientific journals in the northern hemisphere. In 1977 Irving Friedman, a geologist at the University of Pennsylvania who had been working with Brazilian researchers to analyse the isotopes of Amazonian water, warned in a letter to the leading American journal *Science* that cutting down the rainforest could turn the Amazon into an arid savanna like the Sahel of north central Africa. In his article, 'The Amazon Basin, Another Sahel?', he argued that most of the rainfall in the region was recycled from the forests. He wrote, 'The jungle trees with their deep roots act like giant pumps taking water from the water table – often more than 2 meters below the ground surface and transferring it into the air from which it falls again as rain.'[30]

Isotopic analysis of rainfall in the Amazon basin provided more empirical evidence supporting the rainfall recycling thesis. By the early 1960s, scientific techniques for determining the isotopic composition of molecules finally gave scientists the ability to partially trace the movement of water through the entire water cycle. Mass spectrometers allowed scientists to understand the positive and negative charges and thus also the mass of ions within molecules. They found that different isotopes within elements and compound chemicals – for example, oxygen and H_2O (water) – had different masses.[31] (Researchers began to measure oxygen isotopes in water sources starting in the 1930s.[32])

In 1961 the World Meteorological Organization, in conjunction with the International Atomic Energy Agency (IAEA), created the Global Network of Isotopes in Precipitation (GNIP) to monitor and map variations in isotopes globally. That year, the American geochemist Harmon Craig published a seminal paper in *Science* identifying consistent variations in hydrogen and oxygen isotopes in atmospheric moisture, a pattern that is now known as the global meteoric waterline.[33] This pattern of variations has helped scientists to understand the different compositions of the world's waters. Craig's paper focused on the relationship between oxygen-18 isotopes and deuterium, and he discovered that water that evaporates from lakes or snow differs from oceanic water. The two types of evaporated water have different masses. Isotopic analysis began to allow scientists to measure the origins – geographically and in terms of physical process – of precipitation.[34]

Brazilian researchers began to collect isotopic information according to IAEA standards in the 1970s to understand variations in water across the country. At INPA, in Manaus, and at other Brazilian meteorological stations, scientists compiled isotopic data for the Amazon basin by taking samples from rain water, river water, lake water and soil water. They compared the new evidence gathered on the isotopic composition of waters throughout the country with data from GNIP. Among the findings, they found that precipitation in the central and eastern parts of the Amazon basin had a significant enrichment of deuterium. They inferred that the enriched deuterium resulted from recycling, because water elsewhere in the country did not have the same levels. They argued that isotopic evidence 'confirms the importance of the reevaporated moisture in the water balance of the area'.[35] Citing Friedman's 1977 letter in *Science*, they noted that 'the absence of a dense plant cover may have far-reaching effects' on the ecology and climate of the Amazon basin.[36]

In 1984 Brazilian scientists Eneas Salati and Peter Vose published a major review paper that examined the extant literature on rainfall recycling in the Amazon. Salati had led much of Brazil's research from his position as director of the Centro de Energia Nuclear na Agricultura at the University of São Paulo. As the former director of INPA, he had first-hand knowledge of the Amazonian basin from living in the region. The paper pointed to multiple lines of evidence for precipitation recycling in the Amazon. Three lines of evidence – estimates of evapotranspiration, isotopic analysis and studies of vapour flux flows – converged to demonstrate that the Amazon recycled rain. Moreover, they estimated that it took only five and a half days for rain to be recycled in the basin.

These findings fundamentally challenged orthodox hydrological views on tropical and temperate climates. Salati and Vose noted that:

> Hydrologists have been inclined to discount the possibility that changes in land use can affect rainfall . . . However, this is almost certainly a fallacy where large amounts of water are being lost to the system through the greatly increased run-off associated with widespread deforestation and where the existing rainfall regime is greatly dependent on recycling.[37]

They concluded that Amazonian deforestation would cause a considerable amount of localized precipitation decline, and Salati's earlier research had indicated that a 10 to 20 per cent reduction in overall rain could lead to dramatic changes to the Amazon's ecosystem. What's more, such an impact would, the 1984 paper warned, 'be felt in Brazil beyond the immediate area' and affect neighbouring parts of the country.[38]

Research on the Amazon changed thinking about wet tropical forests, but its global significance would not sink in until two decades later with the rise of the supply-side school of thinking. Though the research was convincing, the Amazon had unique features for studying tropical hydrology because of 'its size and the generally unidirectional airflow', which meant the findings there would not always be applicable elsewhere.[39] For these reasons, research done on atmospheric recycling in the Amazon, though accepted by meteorologists and climatologists, did not fundamentally change thinking on precipitation in non-tropical climates (such as temperate) until further developments in climate modelling and empirical findings on precipitation recycling arose in other parts of the world during the 2000s.

Deforestation and Climate: From Regional to Global Concern

In the late 1970s and early 1980s, tropical deforestation increasingly took centre stage in scientific concerns about global warming. Developments in climate modelling in those decades renewed interest in the relationship between forests and climate, especially in the wet tropics. Growing alarms about human-induced climate change, especially the idea that increased emissions of greenhouse gasses could increase global temperature, encouraged researchers in several fields to begin thinking about how forests and other landscapes influenced climate on regional and global scales. In 1979 a National Research Council team led by the American meteorologist Jule Charney (1917–1981) predicted that a doubling of carbon dioxide would likely increase the global temperature by 2°C to 3.5°C, with a potential error of 1.5°C. Advances in modelling and measurement in the 1980s and '90s allowed for researchers to understand diverse forest dynamics – for example, carbon absorption, carbon emissions, ozone emissions, albedo and influence on rain – and to incorporate these processes and data into increasingly sophisticated regional and global climate scenarios.

When global warming became an 'actionable crisis', it helped open the door for the revival of interest in the connections between forests and rainfall.[40] In the late 1970s and early 1980s, scientists began to raise alarms that the loss of trees through tropical deforestation could increase greenhouse gasses and albedo. Whereas earlier theories that linked forests and climate focused on local and regional influences, the rapid growth of data coupled with advances in climate modelling encouraged researchers to begin thinking about how forests could influence global climate.

In 1979 the first World Climate Conference emphasized the importance of forests as a giant store of carbon. In the keynote address, Robert M. White, the conference chairman and former head of the U.S. National Oceanic and Atmospheric Administration (NOAA), told the audience that 'evidence continues to accumulate that . . . the consequent reduction of forests reduces the terrestrial reservoir of carbon and further increases carbon dioxide.'[41] South American contributors called for further research on how much carbon the Amazon stored. Juan Burgos from the University of Buenos Aires noted that there was a lack of data on how forests influence climate but estimates from the Amazon suggested that

Amazonian rainforests stored 10 per cent of the world's carbon and absorbed 1 to 2 per cent of the Earth's atmospheric carbon on a yearly basis.[42]

In the late 1970s and early 1980s, climate models were divided on whether the removal of tropical forests would modify global and regional climate by increasing surface albedo and therefore potentially cool the Earth.[43] The mechanism of greenhouse gas warming is well known today – greenhouse gasses 'trap' ultraviolet radiation in the atmosphere – but relatively few people know about albedo. Albedo refers to the reflectivity of light off a surface. It is lowest in a forest, which absorbs up to 80 per cent of sunlight thus making surface temperatures in them warmer, and is highest in a desert or snowpack, which reflect upwards of 80 per cent of light and thus cool the surface due to the lack of retained solar radiation.

The first alarms about albedo-driven climate change came from West Africa. A devastating famine in the Sahel, caused by a half-decade of drought from 1968 to 1974, gained international attention, and some climate experts attributed the famine to human-induced desertification caused by overgrazing, overpopulation and denuded vegetation.[44] The aforementioned MIT climate researcher Charney, a leader in climate modelling and a forerunner of modern climate research, famously advanced this argument. Charney argued that the denudation of vegetation by grazing and human activity increased albedo, which led to cooling via a loss of radiative energy; the decline in energy ultimately weakened the Hadley cell, a global system of air movement, named after George Hadley (1685–1768), in the tropics driven by the circulation of rising warm air northwards and southwards from the equator, which influenced rain to the Sahel; this process therefore caused a decline in rainfall.[45] His theory hinted at rainfall recycling, but his theory did not focus on precipitation recycling. Charney's work on the Sahel inspired a proliferation of research on albedo and deforestation.

Early research theorized that tropical deforestation would lead to regional and global cooling with corresponding declines in rainfall. A *Nature* paper in 1975 concluded that increased albedo caused by deforestation would reduce surface temperature, especially in the tropics, lower global evaporation and rainfall, weaken the Hadley circulation and cool the middle and upper tropical troposphere.[46]

A controversial paper by the famous science popularizer Carl Sagan and others in 1979 argued that albedo caused by deforestation and other human influences, such as fire, caused desertification, a process which had potentially cooled the Earth over thousands of years. Sagan noted,

> during the past several thousand years the Earth's temperatures could have been depressed by about 1 K [1 degree kelvin, equal to a change of 1 degree Celsius], due primarily to desertification, which might have significantly augmented natural processes in causing the present climate to be about 1 to 2 K cooler than the climatic optimum of several thousand years ago.[47]

Sagan and his collaborators suggested that deforestation in the Amazon 'may even be desirable, as a counterbalance to greenhouse heating of the Earth'. However, they noted, 'it would seem prudent, on an issue of possible global importance, to study its implications in some detail before proceeding unilaterally.'[48] Sagan's view was promptly challenged by modelling that showed how historical deforestation had probably produced almost no climatic effect.[49]

By the early 1980s, evidence and new thinking suggested tropical deforestation could lead to warmer temperatures and less rain.

Climate models and the earliest experiments in the 1980s suggested that any cooling caused by increased surface albedo would be counterbalanced by lower evaporation rates, which would raise temperatures.[50] In the late 1980s, the earliest detailed micrometeorological measurements from the central Amazonian rainforests produced data confirming global models that showed deforestation would lead to a net increase in temperature, because the destruction of forests would lower evaporation (which cools the temperature).[51] This perspective has been confirmed by contemporary research and is recognized widely by researchers working in the field of climate change.[52]

Albedo has remained a subject of interest to climate researchers, but its importance in climate models has declined somewhat due to advances in measurements of other sources of climate change, especially carbon.[53] Albedo came to prominence in the 1970s because of satellite imagery that showed striking changes in landscape caused by humans, but results of research on albedo have been somewhat confusing, owing to factors such as latitude, snow cover, the overall carbon stocks of forests and forest-induced evaporation.[54] The most up-to-date research suggests that deforestation-induced albedo would have opposing outcomes depending on latitude and landscape.[55] Tropical deforestation is seen to produce a net warming, because increased carbon released from the forests reduces evaporative cooling and lowers cloud albedo. The transformation of grasslands to forest might also decrease albedo, thus inducing warming.[56] In the higher northern latitudes, deforestation is believed to produce cooling due to the increase of albedo. Snow cover offsets the warming effect of increased carbon emissions.[57] This means that forests in higher latitudes, such as the boreal forests of Scandinavia, contribute a net warming effect.[58]

Forests became the central focus of global policy discussion in the 1990s due to international concerns about carbon dioxide emissions. In 1992 the Kyoto Protocol promoted the idea that protecting tropical forests from deforestation could help to slow the release of carbon dioxide into the atmosphere and thus lessen predicted climate warming.[59] Environmentalists started to include climate change alongside biodiversity loss as major reasons to stop deforestation. A global programme to arrest deforestation began in 2008 when three organizations within the United Nations (the FAO, UNDP and UNEP) established the Program on Reducing Emissions from Deforestation and Forest Degradation (REDD) to decrease carbon emissions from forests and to sequester airborne carbon stocks. Since then, a considerable amount of research has been done to measure the carbon emissions and storage of forests. Deforestation is estimated to produce approximately 12–20 per cent of yearly global carbon emissions. At the same time, forests absorb upwards of one-third to one-quarter of anthropogenic emissions so they currently absorb more carbon than they emit.[60]

The Re-Emergence of Forests and Climate: Science and Policy Implications

The emergence of the supply-side school of hydrology and meteorology within the past decade reflects advances in regional climate modelling, new ideas in atmospheric physics and more accurate measurement of water molecules through the hydrological cycle. Supply-side scholars posit that forests influence climate and weather at regional and global scales in ways that have not been properly acknowledged by climate policy or modelling. The strongest advocates of supply-side thinking challenge both the carbon-centric priority of forest policy and the demand-side perspective of hydrology.

David Ellison and others write: 'For reasons of sustainability, carbon storage must remain a secondary, though valuable, by-product [of forest preservation].'[61] Supply-side researchers argue that non-carbon climatic influences of forests, including rainfall recycling, should have a more central place in forest and water policies. Researchers believe that it is necessary to preserve forests for their climate-stabilizing role.[62] This viewpoint is influencing governments and other leading non-governmental organizations. A 2012 report for the European Union drafted by researchers at the British Meteorological Office argues that forests 'play a major role in the atmospheric circulation and the water cycle on land and may have a role in mitigating regional climate, desertification and water security problems'.[63]

In a 2018 report on forests and water published by the International Union of Forest Research Organizations (IUFRO), Ellison and a team of researchers argued, 'First, there needs to be some relative agreement that the forest-water relationship should be prioritized over the more common forest-related goals of producing timber and/or sequestering carbon.' They argued that these interactions 'have been almost entirely ignored in the management of global freshwater resources'.[64]

Supply-side hydrology contends that the question of what happens to evapotranspiration, the water that is evaporated or transpired from a plant, has not been sufficiently accounted for by hydrologists or foresters. Demand-side studies were designed to answer a specific question about whether tree planting lowers available water in a catchment, but they did not ask what happened to the 'lost' water because when these studies were created scientists could not trace molecules through the atmosphere. Evapotranspiration is well recognised by demand-side researchers but it is seen as something which exists but is not easily quantifiable at the single catchment scale. Thus, the subject is not seen as a high priority for determining local water-management strategies. Forest hydrology in arid regions, such as South Africa, took this view to the extreme: based on evidence from research stations in the country, planted exotic trees are viewed as water hungry and ecologically destructive introductions to the landscape.[65]

What really happens to moisture after it goes back into the atmosphere? The Dutch researcher Ruud van der Ent conducted an important study that modelled evaporation and precipitation around the world.[66] His research showed that regions with the highest rainfall, such as the Amazon or West Africa, also have the highest rate of evaporation. In other words, forests contributed to evaporated moisture because they transpired water back into the atmosphere. His study suggests that the Amazon, central and East Africa and central parts of Eurasia have the highest proportion of precipitation derived from precipitation recycling. China, Mongolia, Siberia, central and West Africa and the northeast of North America are the main recipients, or 'sinks', of the water. If correct, this modelling suggests that deforestation in Germany and western Russia could lower precipitation in western China. Alternatively, the loss of the Amazon would lead to a drying out of the land in Peru and Uruguay, countries which depend on evaporated moisture.

Van der Ent's study did not look at the potential for modifying precipitation, but the implications are clear: if forests are planted, precipitation would increase in areas downwind. But just how many forests are necessary to create precipitation? That remains an open question, but the larger the scale of the forest the more significant the potential increases in precipitation locally and at downwind sites. Similarly, the loss of forests could lead to even greater decreases in precipitation. Van der Ent's model suggests that a forest-less landscape would have 50 per cent less evapotranspiration than the Earth's current rate – however, this is clearly an estimate. Researchers continue to make new theoretical advances that provide firmer foundations for supply-side thinking and make it more likely that it will become a key government policy for the rest of this century.[67]

One of the more dire projections to emerge from supply-side thinking is that deforestation will lead to an overall decline in regional precipitation. A growing number of studies confirm this view. Research on the Indian monsoon suggests that the summer monsoon has weakened, and thus less rain has fallen on average, as a result of deforestation in India during the nineteenth and twentieth centuries.[68] Scientists have observed similar results in the Congo and Amazon rainforests.[69] Some forests, such as those in northeast India near Assam, have a disproportionate impact on rainfall recycling.[70] Other forests, such as the Blue Nile forests in Ethiopia, exist only because the highland forests capture moisture recycled from atmospheric teleconnections, rivers of moisture flowing through the sky, originating thousands of kilometres away.[71] Trees and forests contribute to moisture recycling systems in both arid and wet regions.

If scientists adopt supply-side thinking, this would likely encourage reforestation, a process that uses natural regeneration to restore forests, and afforestation, the creation of new forests where none existed. As the public becomes more aware that forests play a central role in moisture and temperature regulation, this may increase societal and government interest in tree planting and forest restoration.

Already scholars are modelling areas where large-scale reforestation or afforestation efforts would have the greatest climatic impact.[72] One widely debated model of the global potential for reforestation suggests that there are 4.4 billion hectare (10.8 billion ac) – more than twice the size of the United States, including Alaska – of potential forest to restore or create.[73]

Historical thinking can help to identify and alleviate potential problems that may arise from policy changes. These problems are neither inevitable nor inherently related to supply-side thinking. Supply-side researchers rightly warn about the risks of planting trees where they do not belong.[74] Scientists today have much better knowledge of the hydrological cycle, can access the results of over a hundred years of forestry and ecological research, and understand the importance of protecting biodiversity and the rights of Indigenous people more than at any time in the past. Yet this information is not widely known among the public or even by academics outside of the environmental sciences.

The question of how to plant trees or restore forests is not as simple as it might seem at first. Forests bring innumerable benefits, but they are not a panacea for all environmental issues. Planting the wrong species on the wrong site can create a host of problems, from introducing invasive exotic plants to changing biodiversity in negative ways. In many instances, it will not be viable to plant or sustain trees in many locations and climates.

There are a range of negative consequences that have been associated with poorly designed forestry initiatives, from the loss of biodiversity to the loss of power by local communities when the state or NGOs take over management authority of forests. In the past, the idea that forests increase precipitation reinforced an enthusiasm for all things trees. Tree-planting programmes sprung up even in the most inopportune places – such as in deserts and along streams and rivers in extremely dry climates.

Forests, if planned well, can bring growth to society as well as nature. Walking through or spending time in forests has been shown to improve human health.[75] Being immersed in green for only a few minutes a day can improve a person's mood. Watching Australian rainbow lorikeets flit from branch to branch hunting nectar, seeing fireflies flickering among trees on a late summer evening in southern Missouri and driving down and admiring the Jacaranda-lined streets of Johannesburg: all are moments in which forests and trees improve our lives. Economists, too, can quantify this in terms of money: property values increase in locations surrounded by forests, due to residents' higher satisfaction.[76]

If we wanted to increase the size of our forest cover, where would people make them? Forests can grow in two ways: by people planting them or by natural regeneration. Former forests provide the most obvious places for restoration. In some parts of the world, lands once converted to farmland have now been reclaimed as forests. Forests have grown back in many places, such as in the eastern United States, due to the reforesting of land that was previously used for agricultural purposes. In some cases, an owner or land manager planned this regrowth. During the second half of the twentieth century, farm forestry initiatives around the world encouraged farmers to plant species recommended and they were even supplied, sometimes for free, by governments. These examples provide important case studies for restoration ecology because they reveal what is possible in forest restoration. One key finding from this discipline is that better planning is required to pursue successful forest restoration because often the projects require more money, labour and time to succeed.[77]

Even if humans do not actively influence forest creation and expansion, regrowth occurs spontaneously. In urban and former agricultural areas, the seedbank, that is the seeds found in the local soil, includes a large proportion of introduced species and weeds. Conservation ecologists have expressed alarm about the ecological impact of introduced invasive species that threaten indigenous species, not to mention the invasive species' aesthetic impact on landscape (think the weeds growing from an overgrown, untended plot of urban land). Uncontrolled regeneration does not necessarily help larger ecological restoration efforts, but some view weeds, both introduced and native, as nature's bandage.

Lastly, people can plant trees actively. Tree planting uses up more money, time and labour than natural regeneration, and for that reason is usually reserved for intense economic purposes, such as commercial timber plantations, or for cultural and aesthetic purposes, such as trees planted for streets or private gardens. Creating new forests is extremely labour intensive. The Civilian Conservation Corps, which planted more trees in the 1930s than any other organization in world history, employed more than 3 million people (note: not all were involved in tree planting). Large-scale reforestation efforts have languished in much of the world for a lack of money and workers.

Even if labour can be paid for, many areas are not sufficiently endowed with the climatic and ecological conditions to support forests. The Great Green Wall tree-planting programme in western China highlights the challenges associated with planting in arid environmental conditions. By 2050 newly planted forests are predicted to cover an incredible 400 million hectares (988 million ac) of land with trees. The trees planted in China, a country with a professional forestry service, have an extremely high death rate because they are often planted in poor locations and not looked after. As a result, many scholars question the long-term viability of these plantations.[78] Chinese researchers and officials continue to improve the survival rate of trees, but it remains to be seen how much money or patience there will be for a multi-decadal tree-planting scheme that is not yet returning significant dividends. Research indicates that tree planting in northern China's Loess Plateau, a dry area the size of France that was once fertile, has increased atmospheric moisture, but whether this will generate enough precipitation to change the ecology or pay for the cost of land, labour and materials for the planting remains unanswered.[79]

HISTORIANS AND SCIENTISTS can work together to ensure that rigorous contemporary science rather than enduring popular myth determines policies and public science communication. The belief that forests increase precipitation has endured for more than five hundred years, at times with vast popular appeal. Its intuitive charm allowed it to survive after the Second World War even when scientific consensus rejected it.[80] The strong following that the idea received, and the interest it holds for the human imagination across diverse backgrounds and cultures, may be because of cultural preferences towards forests and trees.[81] Whatever the cause, a lot of public interest in planting trees leads to initiatives of questionable economic merit that also destroy indigenous species and ecosystems. Tree planting has had negative consequences in grasslands as well as open savanna and heath, as there are relatively few trees in these landscapes due to the occurrence of natural fire. The South African botanist William Bond has warned against planting trees in savanna and grassland areas, which he says are two of the most endangered and underappreciated biomes in the world.[82] He points to the World Resource Institute's goal to 'restore' 100 million hectares (247 million ac) in Africa by planting trees as one example where trees are planted in ecologically and hydrologically sensitive areas. Grasslands suffer the worst of urban development because they provide ideal building conditions and because people are not as attached to grasslands as they are to forests. What is worse, the combined effects of human land management, such as from introduced species of trees and weeds, and carbon dioxide fertilization (a process that makes trees grow faster than grass) of trees via climate change have turned once open savanna and grassland ecosystems into more forest-like systems that in some places have turned into closed canopy structures.[83]

A fixation on forests can lead people to look the other way when it comes to losses in other ecosystems, especially those in which people live. Unfortunately, there is evidence that shows preservation in one place may lead to unexpected destruction and consequences in another. For instance, the desire to protect Brazil's rainforests has possibly encouraged the destruction of the diverse Cerrado grasslands.[84] It is easier to convince donors to give money to save rainforests than it is for the protection of grasslands because grasses do not seem to elicit the same level of emotional response.[85]

Some ecologists and social scientists warn about creating a world where forests are more strictly protected than other ecosystem types. Efforts to preserve or conserve forests have ecological and social consequences. People who live in or utilize forests have for the past 150 years faced the prospect of being forcibly removed from their homes or of having their livelihoods curtailed when threatened by state policies to centralize forest management and protect forests.[86] The continued dispossession of people from forests is one of the enduring results of forest set-asides, the reservation of forests which banned people from living in forests, that swept the world starting in the mid-nineteenth century.

All these problems are neither inevitable nor insurmountable, especially if historians and scientists can work together to ensure that forestry policies are informed by proper historical and scientific analysis. New insights from supply-side thinking will continue to reshape contemporary knowledge about the movement of moisture across the entire water cycle. These ideas have the potential to increase biodiversity, generate positive economic outcomes from new managed forests and influence regional precipitation by expanding forests in appropriate areas. They also present a significant challenge to demand-side policies for water and forest management. From its early modern origins to its ultimate rejection in the mid-twentieth century, climatic botany has once again become an important topic of scientific research and environmental policy.

CONCLUSION
INTERPRETING THE HISTORY OF CLIMATE CHANGE

For good reasons, many people are attracted to studying the history of climate change. This might be because they want to discover what we can learn from the past in order to help with our present situation. Some people may want to find lessons from how societies or individuals responded to similar situations in the past, much as scientists during the COVID-19 pandemic studied the Spanish influenza to learn about societal responses to a parallel crisis. Others, including scientists, are fascinated to know how historical ideas resemble or differ from contemporary ones.

We can use history to better interpret the present day by being clear about how we apply historical evidence and reasoning. Readers will inevitably make comparisons and want to better understand continuities between past and present ideas of climatic botany. There are significant intellectual challenges associated with trying to make comparisons or trace continuities between different periods of history.

A comparison effectively says that two things are like each other even when there are obvious differences. One example of comparison would be to say that the desire to protect trees for climatic purposes in the nineteenth century has similarities to contemporary efforts to reduce CO_2. Both hinge on the idea that humans can negatively change the climate, but each focused on different causes and consequences. A continuity refers to the continuation of an idea or phenomenon across two or more periods of time: for example, Humboldt's ideas that forests increased rainfall continued to inspire scientists for multiple decades throughout the world.

Comparisons and continuities are attractive but problematic historical concepts. They are attractive because they help us to bridge the past and present; they are problematic because they are often applied incorrectly. Efforts to make transhistorical generalizations – the concept that ideas or social organization can endure across multiple generations – often do not hold up to rigorous historical scrutiny because of the confirmation bias inherent in the human mind. Simply, humans tend to distort history by projecting our own present-day beliefs into historical events and individuals. We are more likely to agree with people in the past who seemingly held our beliefs and are more likely to disagree with those whose ideas seem contrary to ours. It is important to highlight 'seemingly' in the previous sentence, as often the similarities between ideas centuries apart are superficial.

Any history that focuses on an issue as serious and emotionally gripping as climate change runs the risk of confirming what we already believe by making problematic comparisons and continuities. This must be guarded against. We might interpret pivotal moments, such as Poivre's short-lived forestry reforms, as a continuation of a long, but ultimately unsuccessful struggle by humans to stop destroying nature and the climate at the hands of capitalism and colonialism. In a different vein, critics of climate change might compare the rise, fall and revival of climatic botany with alternative climate theories, such as the return of the ice age, which once had a larger scientific following but is now viewed as a remote possibility by most of the climate research community.

These interpretations are only correct at a superficial level because they focus solely on general similarities, which tend to fall apart upon closer examination. The first interpretation casts Poivre as a radical, yet he not only supported the colonial plantation industry by stealing spices from the East Indies, he was also a Physiocrat, a member of an economic school that advocated free trade (and which preceded Adam Smith). The second interpretation compares two different theories, climatic botany and the ice ages, and fails to explain how new technology dramatically improved the accuracy of measurements with water molecules. Both examples highlight how easy it is for the human mind to make narratives that support our views.

Histories of science, like this book, can be prone to a type of confirmation bias known as Whig history. The term refers to Thomas Babington Macaulay's (1800–1859) portrayal of English history as a story of progress that ultimately led to the creation of the Whig Party. Any history that has an idea of progress in it is in some ways Whiggish. Science as a discipline tends to inherently view the world through a Whig-tinted progressive lens: new ideas and methods are deemed better and more objectively correct than old ones. Scientists like to imagine themselves as part of a progression of ideas – this intergenerational thinking gives our lives meaning and it acknowledges the actions of others in the past, although generally most scientific fields discount the validity of historical ideas. Scientists may stand on the shoulders of giants, but those giants are now proven to be incorrect. In this way, a Whig perspective always has the current generation at the top of the pyramid.

Whig histories of science tend to present the historical development of ideas, be it the idea of evolution or the creation of new typewriters, as a continuous, and natural, development. In secondary school or university, students might read a biology textbook that discusses how Pythagoras invented geometry or Aristotle established the discipline of biology, but they might not properly comprehend these ideas within their local contexts. Historians of science have compellingly shown that there are just as many, if not more, differences than similarities when we examine how people viewed seemingly similar ideas. Rarely does an idea travel throughout time without substantial changes to its meaning to society. Christopher Columbus, for instance, had no inkling of the periodic table of elements nor could he rely on research on evaporation from the 1600s and 1700s, although his views would have agreed with Poivre and Humboldt that forests in tropical regions led to increased precipitation. Today, researchers can measure the movement of molecules through the atmosphere, something naturalists and scientists in the nineteenth century and before could only dream about doing.

The point here is not to reject all connections or comparisons between our ideas today and those in the past. For instance, the idea that forests influenced precipitation and temperature had some commonalities across time and space. Basic theories on forests and climate in the late nineteenth century show striking similarity to the same ideas being discussed today by supply-side scientists. Scientists in the nineteenth century focused on evaporation, transpiration, the origins of moisture and its movement in the atmosphere and the causes of precipitation. Nineteenth-century scientists could not measure the mass or movement of molecules, gain access to upper atmospheric measurements or use computer modelling, so they lacked the ability to validate their ideas. If a Victorian-era forester could read research now about how rainforests induce atmospheric recycling, they would probably find it convincing.

Do contemporary scientific ideas validate various arguments made by the historical subjects of this book? Did Lake Valencia decline due to deforestation or not, as Humboldt argued? Would planting trees to stop the spread of deserts in North Africa work?

Let us take Humboldt as an example. Humboldt popularized the idea that forests generate rainfall and conserve water supplies based on his observations of the declining waters of Lake Valencia in what is today Venezuela.[1] When Humboldt visited, the lake's water had declined after an intense period of deforestation caused by war and rebellion. Much like we see the visual image of melting glaciers as proof of climate warming, so too did naturalists at the time see the drying of the lake as evidence of climate change.

In a general way, Humboldt's argument that forests influence precipitation agrees with key tenets of contemporary hydrological modelling on the relationship between forests and precipitation. Humboldt's research differed because he focused primarily on localized rainfall, that is, rain within a relatively small area, rather than thinking of precipitation at larger regional or global scales. Contemporary research on the origins of precipitation suggests that, for much of the world, the majority of precipitation comes from moisture in the ocean brought to land by low-pressure systems.[2] Forests play an important role in recycling rainfall that comes in from the ocean. For instance, the northern Tibetan Plateau receives the majority of its precipitation from the Atlantic, from water that has been recycled by vegetation across the entire Eurasian continent.[3] Evaporation is responsible for up to 80 per cent of water resources in western China.[4] Cutting down trees, say, in Germany or the Ural Mountains would lead to a decline of rainfall in Tibet.

Aspects of Humboldt's observations – which occurred in a tropical rainforest – do seem to have contemporary parallels. There is a large and growing volume of research on the Amazon and Congo rainforests that shows how deforestation in rainforest ecosystems can change regional rainfall patterns. The Amazon and West African rainforests can be understood as massive recycling systems that continually recirculate moisture.[5] Despite all these parallels, the best scientific evidence suggests that the specific question of why the water in Lake Valencia declined in the early 1800s was because of lower rainfall due to variations in the El Niño-Southern Oscillation (ENSO).[6]

The Great Green Wall tree-planting programme in the Sahel run by the African Union is partly inspired by Baker's idea of a green wall, which owes many of its origins to Stebbing's idea of creating a wall of forests to stop the expansion of the Sahara. The Great Green Wall runs through 22 countries and is 8,000 kilometres (4,970 mi.) long. Today, the Wall has a climatic focus in terms of carbon sequestration, but it was founded primarily to stop desertification, create employment, improve food security and maintain soil fertility. Stebbing and Baker's belief that forests generated precipitation is not entirely without merit. Research on rainfall in the arid Sahel suggests that 30 to 40 per cent of the region's rainfall comes from recycling.[7] This research suggests that all earlier efforts to change the climate would have succeeded more if trees had been placed upwind, where evapotranspiration is higher. The winds in the inland parts of West Africa come from northeasterly trade winds coming in from eastern and northern Africa. Trees should be planted upwind from where people want the weather to be changed, rather than at the site itself. Ironically, early foresters may have been right about one aspect of climatic botany, but they planted trees in the wrong place. To increase rain in the Sahel, tree planting should be more focused in East Africa because prevailing winds come off the Indian Ocean into central Africa.

Perhaps the most salient parallel between the past and present comes not from scientific ideas but from our shared environmental anxieties. Like Humboldt, scientists today who worry about the future climate of the world warn that deforestation can cause dramatic fluctuations in precipitation. Humboldt's ideas helped spark the first global environmental movement, which influenced forest management on every inhabited continent. Today we once again hear the argument that forests matter for climate, except this time it focuses on regional and global scales. Supply-side hydrologists argue that the contemporary climate-centric focus on carbon has obscured the important role of water and vegetation, especially forests. If correct, this theory opens new avenues for understanding the dynamics that shape regional climate, and it creates opportunities to rethink how forests are planted within landscapes.

The tendency to judge the past by contemporary standards raises another problem, which comes in the form of a paradox and can be expressed something like this: history offers many parallels but few lessons. Perhaps the best lesson we can draw from this history is the value of remaining open-minded scientifically while at the same time making policy based on the best available evidence. This is how we make meaning from the fact that scientific views on forests and climate shifted considerably over the past centuries. There has never been a single consensus on the question of forests and climate. The supply-side view has an increasingly larger body of evidence to warrant reviewing forest and catchment policies. Still, the demand-side perspective reflects the reality of being a manager of a single catchment – in that instance your goal is not to maximize downwind rain but to conserve existing water within the catchment. Scenarios of future climate change suggest that rain patterns may shift, in some cases dramatically, and it is necessary for society to start thinking about how forest restoration can alleviate the worst of these changes and even possibly to modify precipitation in ways that are positive to humans and nature.

The story of the rise, fall and revival of climatic botany told in this book demonstrates how ideas of forests and climate ebbed and flowed in popularity across the last three centuries. The theory of climatic botany has endured for many centuries and is unlikely to go away in our lifetime. We are witnessing a revival of the link between forests and precipitation. Will these ideas rise only to fall once again to some yet unknown scientific paradigm? Only time will tell. For now, there are many exciting reasons to believe that modern science holds the keys to better managing our forest and water resources for the good of humans and nature. Importantly, humanity's ability to appropriately balance the interests of all people with nature will be determined not only by science, but also by our ability to learn from history.

REFERENCES

Introduction: The Forgotten History of Climatic Botany

1 See Richard Grove, *Green Imperialism: Colonial Expansion, Tropical Island Edens and the Origins of Environmentalism, 1600–1860* (Cambridge, 1995), pp. 196–7. *Green Imperialism* was an important book that opened up historical research on forests and climate change, especially in the early modern era. It remains one of the most detailed studies of ideas of climatic botany in the early modern period. The book has been challenged for its emphasis on non-European examples and agents. For an important recent critique of Grove's thesis that environmentalism emerged primarily in a colonial rather than a European context, see Fabien Locher and Jean-Baptiste Fressoz, *Les Révoltes du ciel: Une histoire du changement climatique, XVe–XXe siècle* (Paris, 2020).

2 Ibid., p. 222.

3 Cited ibid., pp. 196–7; Pierre Poivre, *Travels of a Philosopher* (Dublin, 1770), p. 30.

4 Cited in Fabien Locher and Jean-Baptiste Fressoz, 'Modernity's Frail Climate: A Climate History of Environmental Reflexivity', *Critical Inquiry*, XXXVIII/3 (2012), pp. 379–98, 582.

5 For the origins of this debate, see Fredrik Albritton Jonsson, 'Rival Ecologies of Global Commerce: Adam Smith and the Natural Historians', *American Historical Review*, CXV (2010), pp. 1342–63.

6 Gregory Barton, *Empire Forestry and the Origins of Environmentalism* (Cambridge, 2002).

7 Meredith McKittrick, 'Theories of "Reprecipitation" and Climate Change in the Settler Colonial World', *History of Meteorology*, VIII (2017), p. 75. McKittrick's excellent article notes that the forest–climate question has endured but her study does not focus on the period from 1970 to the present.

8 David Ellison et al., 'Trees, Forests and Water: Cool Insights for a Hot World', *Global Environmental Change*, XLIII (2017), pp. 51–61, 52.

9 Ruud J. van der Ent et al., 'Origin and Fate of Atmospheric Moisture over Continents', *Water Resources Research*, XLVI/9 (2010), pp. 1–12 (pp. 7–8).

10 Brett Bennett and Frederick Kruger, *Forestry and Water Conservation in South Africa: History, Science and Policy* (Canberra, 2015).

11 Ellison et al., 'Trees, Forests and Water', p. 51.

12 Michael Sanderson et al., 'Influences of EU Forests on Weather Patterns: Final Report', *Meteorological Office Hadley Centre* (2012), pp. 1–32 (p. 1).

13 Most of these histories focus on the story of colonial forestry and botany of former colonies. Most work has been done on the history of forestry in India, but there are well-developed literatures on southern Africa, North Africa, Australia and New Zealand, the United States and more. A brief list of key works includes Grove, *Green Imperialism*; Barton, *Empire Forestry*; Diana K. Davis, *Resurrecting the Granary of Rome: Environmental History and French Colonial Expansion in North Africa* (Athens, OH, 2007); James Beattie, *Empire and Environmental Anxiety: Health, Science, Art and Conservation in South Asia and Australasia, 1800–1920* (New York, 2011); Kalyanakrishnan Sivaramakrishnan, *Modern Forests: Statemaking and Environmental Change in Colonial Eastern India* (Stanford, CA, 1999); Ramchandra Guha, *The Unquiet Woods: Ecological Change and Peasant Resistance in the Himalaya* (Berkeley, CA, 1989).

14 For climate histories that discuss climatic botany, see James Fleming, *Historical Perspectives on Climate Change* (New York, 1998), Ch. 2; Paul Warde, Libby Robin and Sverker Sörlin, *The Environment: A History of the Idea* (Baltimore, MD, 2018), pp. 33, 105–8. The emphasis on carbon dioxide has spawned a large and exciting literature not discussed in this book. See Spencer Weart, *The Discovery of Global Warming* (Cambridge, MA, 2003); Wolfgang Behringer, *A Cultural History of Climate* (London, 2010); Mark Carey, *In the Shadow of Melting Glaciers: Climate Change and Andean Society* (New York, 2010); Paul N. Edwards, *A Vast Machine: Computer Models, Climate Data, and the Warming of Global Politics* (Cambridge, MA, 2013); Andreas Malm, 'The Origins of Fossil Capital: From Water to Steam in the British Cotton Industry', *Historical Materialism*, XXI/1 (2013), pp. 15–68; J. R. McNeil and Peter Engelke, *The Great Acceleration: An Environmental History of the Anthropocene since 1945* (Cambridge, MA, 2016); and Richard Rhodes, *Energy: A Human History* (New York, 2018).

15 Rohan D' Souza, 'Supply-Side Hydrology in India: The Last Gasp', *Economic and Political Weekly*, XXXVIII/36 (2003), pp. 3785–90.

16 Brett Bennett, *Plantations and Protected Areas: A Global History of Forest Management* (Cambridge, MA, 2015), p. 22.

17 Diana K. Davis, *The Arid Lands: History, Power, Knowledge* (Cambridge, MA, and London, 2016), p. 55; Clarence Glacken, *Traces on the Rhodian Shore: Nature and Culture in Western Thought from Ancient Times to the End of the Eighteenth Century* (Berkeley and Los Angeles, CA, and London, 1976), pp. 129–30.

18 Reflexivity does not here denote modern ecological or environmental thought, but it refers to the human ability to recognize that human impacts on nature can be detrimental to livelihood. See the work by Locher and Fressoz, 'Modernity's Frail Climate' and *Les Révoltes du ciel*. For discussions from ancient Greece and Rome, see Glacken, *Traces on the Rhodian Shore*, pp. 118, 131–6, 145.

19 Glacken, *Traces on the Rhodian Shore*, pp. 131–2.

20 Ibid., p. 311.
21 Lydia Barnett, *After the Flood: Imagining the Global Environment in Early Modern Europe* (Baltimore, MD, 2019).
22 Climate change and extinction both implicated humans in altering God's world.
23 Locher and Fressoz, 'Modernity's Frail Climate', p. 582.
24 Davis, *Resurrecting the Granary of Rome*.
25 See Philip Curtin, 'The End of the "White Man's Grave"? Nineteenth-Century Mortality in West Africa', *Journal of Interdisciplinary History*, XXI/1 (1990), pp. 63–88.
26 Many scholars, such as Richard Grove, use the term 'desiccation', but this only describes one part of the process of climatic botany. Desiccation refers to the process whereby a landscape and climate dries and loses moisture throughout the entire hydrological cycle. Climatic botany also takes into account efforts to increase precipitation and overall moisture availability in soil, vegetation and atmosphere.
27 Grove, *Green Imperialism*, p. 30. The rest of the paragraph refers to research from Grove's *Green Imperialism*.

1 Redeeming the New World

1 The wider history of American ideas of climate in the early modern period has been covered by a number of authors. Some key works include Anya Zilberstein, *A Temperate Empire: Making Climate Change in Early America* (Oxford, 2016); Lee Alan Dugatkin, *Mr Jefferson and the Giant Moose: Natural History in Early America* (Chicago, IL, 2009); James Fleming, *Historical Perspectives on Climate Change* (Oxford, 1998), Ch. 2; Antonello Gerbi, *The Dispute of the New World: The History of a Polemic, 1750–1900* (Pittsburgh, PA, 1973).
2 Gregory Cushman, 'Humboldtian Science, Creole Meteorology, and the Discovery of Human-Caused Climate Change in Northern South America', *Osiris*, XXVI (2011), pp. 19–44 (pp. 26–7).
3 Thomas Jefferson, *Notes on the State of Virginia* (Philadelphia, PA, 1801), p. 79. For a succinct discussion of Jefferson and Buffon's debate, see Gordon S. Wood, 'America's First Climate Debate: Thomas Jefferson Questioned the Science of European Doomsayers', *American History*, XL/6 (2010), pp. 58–63.
4 Thomas Jefferson, 'Bernard Lacépède', in *Letters of the Lewis and Clark Expedition with Related Documents, 1783–1854*, ed. Donald Jackson, 2nd edn (Urbana, IL, 1978), pp. 15–60.
5 Jefferson, *Notes on the State of Virginia*, p. 114.
6 Cushman, 'Humboldtian Science', pp. 26–7.
7 C. A. Bayly, *The Birth of the Modern World, 1780–1914* (Oxford, 2004), pp. 27–36.
8 The dawning of the nineteenth century ushered in the 'health transitions'. See James C. Riley, *Rising Life Expectancy: A Global History* (Cambridge, 2001).

9 Katherine Park and Lorraine Daston, 'Introduction: The Age of the New', in *The Cambridge History of Science*, vol. III, ed. Park and Daston (Cambridge, 2006), pp. 1–17.
10 See Fredrik Albritton Jonsson, *Enlightenment's Frontier: The Scottish Highlands and the Origins of Environmentalism* (New Haven, CT, 2013); Richard Drayton, *Nature's Government: Science, Imperial Britain, and the 'Improvement' of the World* (New Haven, CT, 2000); John Gascoigne, *Science in the Service of Empire: Joseph Banks, the British State, and the Uses of Science in the Age of Revolution* (Cambridge, 1998); John Gascoigne, *Joseph Banks and the English Enlightenment: Useful Knowledge and Polite Culture* (Cambridge, 1994).
11 See Jonsson, *Enlightenment's Frontier*.
12 David Hume, *Dialogues concerning Natural Religion* (London, 1779), p. 211.
13 See Arthur O. Lovejoy, *The Great Chain of Being* (Cambridge, MA, 1964).
14 For a discussion on the differences between how Christianity, Hinduism and Buddhism could influence environmental ethics, see Gregory Barton, 'Abolishing the East: The Dated Nature of Orientalism in the Definition and Ethical Analysis of the Hindu Faith', *Comparative Studies in South Asia, Africa, Middle East*, XXIX/2 (2009), pp. 281–90.
15 Lynn White Jr, 'The Historical Roots of Our Ecological Crisis', *Science New Series*, CLV/3767, pp. 1203–7.
16 Gregory Barton, *Empire Forestry and the Origins of Environmentalism* (Cambridge, 2002), p. 165. Other leading environmental figures such as John Muir were shaped by Protestantism that had become secularized. Donald Worster, *A Passion for Nature: The Life of John Muir* (New York, 2008), pp. 209, 415. There were few clear-cut 'religious' or 'secular' strains of environmental thought prior to the twentieth century because religious beliefs and secular notions mixed together.
17 Peter J. Bowler, *The Norton History of the Environmental Sciences* (New York, 1993), pp. 186–7.
18 Cushman, 'Humboldtian Science', pp. 30, 34–5. Also, see Dugatkin, *Mr Jefferson and the Giant Moose*, p. 114.
19 See Mark V. Barrow Jr, *Nature's Ghosts: Confronting Extinction from the Age of Jefferson to the Age of Ecology* (Chicago, IL, 2009).
20 Erasmus Darwin, *The Botanic Garden: A Poem in Two Parts; containing the economy of vegetation and the loves of the plants; with philosophical notes* (Cambridge, MA, 1825), p. 139.
21 Peter J. Bowler, *Evolution: The History of an Idea* (Berkeley and Los Angeles, CA, 2003), pp. 108–15.
22 Comte de Buffon, *Histoire naturelle, générale et particulière*, XVIII/122 (1764), p. 154; Jefferson, *Notes*, p. 89. Jean-Jacques Rousseau is frequently attributed with describing the people of the New World as 'noble savages', but this is, in fact, a misattribution based on Voltaire's criticism of the supposed belief (which Rousseau rejected) of primitivism. See Terry Jay Ellingson, *The Myth of the Noble Savage* (Berkeley and Los Angeles, CA, 2001).

23 John Woodward, 'Some Thoughts and Experiments concerning Vegetation', *Miscellanea Curiosa*, I (1708), p. 220. As cited in Kenneth Thompson, 'Forests and Climate Change in America: Some Early Views', *Climatic Change*, III/1 (1980), pp. 47–64 (p. 48).
24 Edward Gibbon, *The History of the Decline and Fall of the Roman Empire* [1776] (London, 1810), p. 346.
25 Charles Montesquieu, *Pensées et fragments inédits de Montesquieu*, ed. Gaston de Montesquieu (Bordeaux, 1899–1901); David Hume, 'Of the Populousness of Ancient Nations', in *Essays, Moral, Political and Literary*, ed. T. H. Green and T. H. Grose (London, 1882), pp. 356–60.
26 Alexander Porteous, *Forest Folklore, Mythology, and Romance* (London, 1928), pp. 84–148. As cited in Michael Williams, *Americans and Their Forests: A Historical Geography* (Cambridge, 1992), p. 10.
27 Clarence Glacken, *Traces on the Rhodian Shore: Nature and Culture in Western Thought from Ancient Times to the End of the Eighteenth Century* (Berkeley, CA, 1967).
28 For fears of American forests see Williams, *Americans and Their Forests*, pp. 10–14.
29 See Karen Ordahl Kupperman, 'The Puzzle of the American Climate in the Early Colonial Period', *American Historical Review*, LXXXVII/5 (1982), pp. 1262–89.
30 William Cronin, *Changes in the Land: Indians, Colonists, and the Ecology of New England* (New York, 1983), pp. 108–26.
31 Kupperman, 'The Puzzle of the American Climate in the Early Colonial Period', p. 1266.
32 Fleming, *Historical Perspectives on Climate Change*, p. 24.
33 Dugatkin, *Mr Jefferson and the Giant Moose*, p. 31.
34 Jefferson, *Notes*, pp. 70–71.
35 Ibid., p. 71.
36 Ibid., p. 134.
37 Ibid.
38 Kevin Hayes, *The Road to Monticello: The Life and Mind of Thomas Jefferson* (New York, 2008), pp. 302–4.
39 Charles Wood, 'Environmental Hazards, Eighteenth-Century Style', in *Old World, New World: America and Europe in the Age of Jefferson*, ed. Leonard J. Sadosky, Peter Nicolaisen, Peter S. Onuf and Andrew J. O'Shaughnessy (Monticello, 2010), pp. 15–31 (p. 16).
40 Ibid., pp. 22–3. See Dugatkin, *Mr Jefferson and the Giant Moose.*
41 From William Currie, *An Historical Account of the Climates and Diseases of the United States of America* (Philadelphia, PA, 1792). Cited in Jan Golinski, 'Debating the Atmospheric Constitution: Yellow Fever and the American Climate', *Eighteenth-Century Studies*, XLIX/2 (2016), pp. 149–65 (p. 152).
42 Samuel Williams, *The Natural and Civil History of Vermont*, vol. II (Burlington, VT, 1809), p. 75.

43 Ibid., p. 70.
44 Noah Webster, *A Dissertation on the Supposed Change in the Temperature of Winter: Read before the Connecticut Academy of Arts and Sciences, 1799* (New Haven, CT, 1799), p. 2.
45 Ibid., p. 40.

2 When Climate Change Became Bad

1 Michael Williams, *Deforesting the Earth: From Prehistory to Global Crisis* (Chicago, IL, 2006), p. 264.
2 Ibid., p. 351.
3 See Nicolaas A. Rupke, *Alexander von Humboldt: A Metabiography* (Chicago, IL, 2008).
4 See Humboldt to David Friedlander (11 April 1799) in *Die Jugendbriefe Alexander von Humboldts*, ed. Ilse Jhan and Fritz G. Lange (Berlin, 1973), p. 657. As cited in Michael Dettelbach, 'Humboldtian Science', in *Cultures of Natural History*, ed. Nicholas Jardine, James A. Secord and Emma C. Spray (Cambridge, 1996), p. 290.
5 Ibid.
6 Gregory Cushman, 'Humboldtian Science, Creole Meteorology, and the Discovery of Human-Caused Climate Change in Northern South America', *Osiris*, XXVI (2011), pp. 19–44 (p. 35).
7 Andrea Wulf, *The Invention of Nature: The Adventures of Alexander von Humboldt: The Lost Hero of Science* (London, 2015), p. 58.
8 Laura Dassow Walls, *The Passage to Cosmos: Alexander von Humboldt and the Shaping of America* (Chicago, IL, 2009), p. 66.
9 Paul Ward, *The Invention of Sustainability: Nature and Destiny, c. 1500–1870* (Cambridge, 2018).
10 Hugh Williamson, *Observations on the Climate in Different Parts of America* (New York, 1811), pp. 24–5.
11 Alexander von Humboldt, *Equinoctial Regions of America*, vol. II (Frankfurt, 2019), p. 10.
12 Alexander von Humboldt and Aimé Bonpland, *Personal Narrative of Travel to the Equinoctial Regions of the New Continent during the Years, 1799–1804* (London, 1819), vol. IV, p. 143.
13 Ibid., p. 10.
14 F. J. de Pons, *Travels in South America, during the Years 1801, 1802, 1803, and 1804* (London, 1807), vol. I, p. 74.
15 Ibid., p. 75.
16 Alexander von Humboldt, *Voyage aux régions équinoxiales du Nouveau Continent, fait en 1799, 1800, 1802, 1803 et 1804* (Paris, 1816–31).
17 *The Monthly Anthology and Boston Review* (Boston, MA, 1807), pp. iv, 447.
18 Humboldt and Bonpland, *Personal Narrative*, pp. 142–3.
19 Josiah Conder, *A Popular Description of Colombia: Geographical, Historical, and Topographical* (London, 1825), p. 189.

20 *Supplement to the Fourth, Fifth and Six Editions of the Encyclopaedia Britannica with Preliminary Dissertations on the History of the Sciences* (Edinburgh, 1824), pp. ii, 613.
21 Gregory T. Cushman, *Guano and the Opening of the Pacific World: A Global Ecological History* (Cambridge, 2013), p. 33. Also see Richard Grove, *Green Imperialism: Colonial Expansion, Tropical Island Edens and the Origins of Environmentalism, 1600–1860* (Cambridge, 1995), pp. 376–8.
22 Jean B. Boussingault, *Rural Economy, in Its Relations with Chemistry, Physics, and Meteorology* (New York, 1845), p. 499.
23 Ibid.
24 Conder, *A Popular Description of Colombia*, p. 189.
25 Thomas Rogers, *The Deepest Wounds: A Labor and Environmental History of Sugar in Northeast Brazil* (Chapel Hill, NC, 2010), p. 49.
26 For a discussion on climate change in the public sphere, see Mike Hulme, *Why We Disagree about Climate Change: Understanding Controversy, Inaction and Opportunity* (Cambridge, 2009).
27 William Paley, *Paley's Natural Theology with Illustrative Notes, &c*, vol. II (New York, 1842), p. 154.
28 Fredrik Albritton Jonsson, *Enlightenment's Frontier: The Scottish Highlands and the Origins of Environmentalism* (New Haven, CT, 2013), pp. 49, 131–4.
29 Cited ibid., p. 244. From William Aiton, *A Treatise on the Origin, Qualities, and Cultivation of Moss-Earth* (Glasgow, 1805), title page.
30 Karl Marx's ideas of the environment allowed for human-induced climate change, but many classical economists downplayed the idea of environmental change. See Kohei Saito, *Marx in the Anthropocene towards the Idea of Degrowth Communism* (Cambridge, 2023).
31 Thomas Malthus, *An Essay on the Principle of Population* (London, 1798).
32 Fredrik Albritton Jonsson, 'Rival Ecologies of Global Commerce: Adam Smith and the Natural Historians', *American Historical Review*, CXV (2010), pp. 1342–63.

3 Stopping Climate Change in British India

1 *Allen's Indian Mail, and Register of Intelligence for British and Foreign India and China, and All Parts of the East* (London, 1852), pp. x, 636.
2 Sir William Lee-Warner, *The Life of the Marquis of Dalhousie* (London and New York, 1904), p. 403.
3 Ibid., p. viii.
4 Gregory Barton, *Empire Forestry and the Origins of Environmentalism* (Cambridge, 2002).
5 Leonard Huxley, ed., *Life and Letters of Sir Joseph Dalton Hooker* (Cambridge, 1918), vol. II, p. 232.
6 Ibid.
7 A part of Humboldt's letter to Hooker was published in the *London Journal of Botany*, VI (1847), pp. 604–7. For Hooker's relationship with Humboldt,

see Moritz von Brescius, *German Science in the Age of Empire* (Cambridge, 2018), pp. 42–3.

8 Richard Grove, *Green Imperialism: Colonial Expansion, Tropical Island Edens and the Origins of Environmentalism, 1600–1860* (Cambridge, 1996), p. 364.

9 Joseph Hooker, 'Joseph Hooker to Charles Darwin 20 February–16 March 1848', *Darwin Correspondence Project*, www.darwinproject.ac.uk, 20 February 1848.

10 Huxley, *Life and Letters of Sir Joseph Dalton Hooker*, p. 232.

11 Hooker, 'Joseph Hooker to Charles Darwin 20 February–16 March 1848'.

12 Joseph Hooker, 'Joseph D. Hooker to Charles R. Darwin', *Darwin Correspondence Project*, www.darwinproject.ac.uk, 4 March 1848.

13 Joseph Hooker, *Himalayan Journals; or, Notes of a Naturalist in Bengal, the Sikkim and Nepal Himalayas, the Khasia Mountains, &c. in Two Volumes* (London, 1854).

14 'Joseph Hooker to Charles Darwin, 20 February–16 March 1848', in *The Correspondence of Charles Darwin: 1847–1850*, ed. Frederick Burkhardt and Sydney Smith (Cambridge, 1989), vol. IV, p. 116.

15 Grove, *Green Imperialism*, pp. 396–8.

16 George Robert Gleig, *The Life of Sir Thomas Munro, Late Governor of Madras with Correspondence and Private Papers*, vol. II (London, 1830), p. 350.

17 Ibid., p. 354.

18 Hugh Falconer, *Selections from the Record of the Bengal Government, No. 9, Report on the Teak Forests of the Tenaserim Provinces* (Calcutta, 1852), p. 204.

19 See, for example, Barton, *Empire Forestry*, pp. 46–7; Kalyanakrishnan Sivaramakrishnan, *Modern Forests: Statemaking and Environmental Change in Colonial Eastern India* (Stanford, CA, 1999), p. 204.

20 Grove, *Green Imperialism*, p. 397.

21 Gleig, *The Life of Sir Thomas Munro*, p. 360. Richard Grove notes the parallels between Munro's comments and later criticisms of the Indian Forest Service. See Grove, *Green Imperialism*, p. 398.

22 Grove, *Green Imperialism*, p. 359.

23 Ibid., p. 397.

24 Hugh Cleghorn, *The Forests and Gardens of South India* (London, 1861), p. 9.

25 Hugh Cleghorn, Forbes Royle and R. Strachey, 'Report of the Committee Appointed by the British Association to consider the probable effects in an economical and physical point of view on the Destruction of Tropical Forests', in *Report of the Twenty-First Meeting of the British Association for the Advancement of Science* (London, 1852), pp. 78–102.

26 See the quote from Grove, *Green Imperialism*, p. 455. Original citation from J. L. Stuart, ed., *Select Papers of the Agri-Horticultural Society of the Punjab from Its Commencement to 1862* (Lahore, 1865), pp. 1–5.

27 Barton, *Empire Forestry*, pp. 51–4.

28 Diana K. Davis, *Resurrecting the Granary of Rome: Environmental History and French Colonial Expansion in North Africa* (Athens, OH, 2007), p. 79.

The French North African experience paralleled and in some respects guided the Indian story.

29 See especially Indra Munshi Saladanha, 'Colonialism and Professionalism: A German Forester in India', *Environment and History: South Asia Special Issue*, II/2 (1996), pp. 195–219.

30 See Greg Barton and Brett M. Bennett, 'Forestry as Foreign Policy: Anglo-Siamese Relations and the Origins of Britain's Informal Empire in the Teak Forests of Northern Siam, 1883–1925', *Itinerario*, XXXIV/2 (2010), pp. 65–86; Greg Barton and Brett M. Bennett, 'A Case Study in the Environmental History of Gentlemanly Capitalism: The Battle between Gentlemen Teak Merchants and State Foresters in Burma and Siam, 1827–1901', in *Africa, Empire, and Globalization: Essays in Honor of A. G. Hopkins*, ed. Toyin Falola and Emily Brownell (Durham, NC, 2011), pp. 317–32.

31 G. F. Pearson, 'Report on the Administration of the Forest Department in the Several Provinces under the Government of India, 1870–71. With Appendices', *Indian Office Records, British Library*, XXIV/1242 (1871), p. 49.

32 Some works on the imperial dimensions of this debate include Barton, *Empire Forestry*; Grove, *Green Imperialism*; James Beattie, *Empire and Environmental Anxiety: Health, Science, Art and Conservation in South Asia and Australasia, 1800–1920* (New York, 2011); and Davis, *Resurrecting the Granary of Rome*.

33 The idea that the French could restore ancient Roman climatic conditions, which supposedly had more rainfall, influenced this thinking. See Diana K. Davis, 'Restoring Roman Nature: French Identity and North African Environmental History', in *Environmental Imaginaries of the Middle East and North Africa*, ed. Diana K. Davis and Edmund Burke III (Athens, OH, 2011).

34 Davis, *Resurrecting the Granary of Rome*, p. 107.

35 Brett Bennett, 'A Network Approach to the Origins of Forestry Education in India, 1855–1885', in *Science and Empire*, ed. Joseph Hodge and Brett Bennett (Basingstoke, 2011), pp. 68–88 (pp. 75–80).

36 Nicola McLelland, 'The History of Language Learning and Teaching in Britain', *Language Learning Journal*, XLVI/1 (2018), pp. 6–16 (p. 7).

37 William Schlich, *Schlich's Manual of Forestry*, vol. I: *Forest Policy in the British Empire* (London, 1922), p. 67.

38 Barton, *Empire Forestry*, p. 1.

39 Brett Bennett, *Plantations and Protected Areas: A Global History of Forest Management* (Cambridge, MA, 2015), pp. 33–4.

40 For the United States, see Michael Williams, *Americans and Their Forests: A Historical Geography* (Cambridge, 1989), pp. 376–87.

41 The specific date varies according to location. In the United States, the idea began to decline in the 1890s, whereas in India the idea remained potent for much longer. See ibid., p. 287. See the next chapter for further detail.

42 Sir William Schlich, trans. B. H. Baden-Powell, 'The Influence of Forests on Rainfall', *The Indian Forester*, XIV (1888), p. 18.

43 Berthold Ribbentrop, *Forestry in British India* (Calcutta, 1900), pp. 37–8.
44 Ibid.

4 The Evaporation of the Forest–Climate Question

1 For the full speech and a historical contextualization see Jamie Lewis, 'Theodore Roosevelt's Cautionary Tale', *Forest History Today* (2005), pp. 53–7.
2 Robert DeCourcy Ward, 'The Influence of Forests upon Climate', *Popular Science Monthly*, LXXXII (1913), p. 314.
3 *Hearings before the Committee on Agriculture during the Second Session of the Sixty-First Congress* (1909), vol. III, p. 90.
4 Ian Tyrrell, *Crisis of the Wasteful Nation: Empire and Conservation in Theodore Roosevalt's America* (Chicago, IL, 2015), p. 75.
5 See Gordan Dodds, 'The Stream-Flow Controversy: A Conservation Turning Point', *American History*, LVI/1 (1961), pp. 59–69, and Lincoln Bramwell, '1911 Weeks Act: The Legislation That Nationalized the U.S. Forest Service', *Journal of Energy and Natural Resources Law*, XXX/3 (2012), pp. 325–36.
6 Gordon B. Dodds, *Hiram Martin Chittenden: His Public Career* (Lexington, KY, 1973), p. 161.
7 Ibid., p. 138.
8 Hiram Martin Chittenden, *The Yellowstone National Park: Historical and Descriptive* (Cincinnati, OH, 1895), p. 188.
9 See Hiram Martin Chittenden, 'Forests and Reservoirs in Their Relation to Stream Flow with Particular Reference to Navigable Rivers', *Transactions of the American Society of Civil Engineers*, LXII/1 (1908).
10 Ibid., p. 268.
11 *The Foreign Relations of the United States with the Annual Message of the President Transmitted to Congress December 8, 1908* (Washington, DC, 1912), p. xxxiv.
12 George Perkins Marsh, *Man and Nature; or, Physical Geography as Modified by Human Actions* (New York, 1864), p. 189.
13 Ibid., p. 3.
14 Paul N. Edwards, *A Vast Machine: Computer Models, Climate Data, and the Warming of Global Politics* (Cambridge, MA, 2013), p. 69.
15 Dodds, *Hiram Martin Chittenden*, p. 68.
16 See Deborah R. Coen, *Climate in Motion: Science, Empire and the Problem of Scale* (Chicago, IL, 2018).
17 James Fleming, *Historical Perspectives on Climate Change* (Oxford, 1998), p. 53.
18 Ibid., p. 51.
19 Edwards, *A Vast Machine*, p. 62.
20 Julius von Hann, *Handbook of Climatology*, vol. I (London, 1903), p. 1. Italics from original quote.
21 Ibid.
22 Frederik Nebeker, *Calculating the Weather: Meteorology in the 20th Century*, 1st edn (San Diego, CA, 1995), p. 51.
23 Malcolm Walker, *History of the Meteorological Office* (Cambridge, 2011), p. 15.

24 See Jim Burton, 'Robert FitzRoy and the Early History of the Meteorological Office', *British Journal for the History of Science*, XIX/2 (1986), pp. 147–76.
25 Ibid.
26 'Admiral Fitzroy's Weather Book', *The Spectator*, https://archive.spectator.co.uk, 7 February 1863.
27 Anon, 'Official Daily Weather Reports', *Mechanics Magazine* (1867), pp. 77–8 (p. 77).
28 Coen, *Climate in Motion*, p. 248.
29 Fleming, *Historical Perspectives on Climate Change*, p. 51.
30 Cleveland Abbe, 'Is Our Climate Changing?', *Forum*, XI (1889), pp. 678–88 (p. 689).
31 Willis Moore, *A Report on the Influence of Forests on Climate and Floods* (Washington, DC, 1910), p. 7.
32 The article was circulated in other magazines, including *Mining and Scientific Press*, *Engineering News*, *American Railway Journal* and *Engineering Record News*.
33 Jan Golinski, *British Weather and the Climate of Enlightenment* (Chicago, IL, 2007), p. 115.
34 See Richard Hölzl, 'Historicizing Sustainability: German Scientific Forestry in the Eighteenth and Nineteenth Centuries', *Science as Culture*, XIX (2010), pp. 431–60.
35 Ibid., pp. 5–6.
36 See for instance Berthold Ribbentrop, *Forestry in British India* (New Delhi, 1989), p. 55.
37 Coen, *Climate in Motion*, p. 248.
38 See Aleksandr Ivanovich Woeikof, *On the Influence of Forests upon Climate* (Whitefish, MT, 2010).
39 Ibid., p. 26.
40 H. A. Hazen, 'Forests and Rainfall', *Engineering News* (1898), pp. 5–6.
41 Ward, 'The Influence of Forests', p. 323.
42 Adolphus Greely, *American Weather: A Popular Exposition of the Phenomena of the Weather, Including Chapters on Hot and Cold Waves, Blizzards, Hailstorms and Tornadoes, Etc.* (New York, 1888), p. 157. H. P. Blanford, *Rainfall of India* (Calcutta, 1886), pp. 33–47.
43 India Meteorological Department, *Report on the Administration of the Meteorological Department of the Government of India, 1885–6* (Calcutta, 1887), p. 9.
44 Hann, *Handbook of Climatology*, p. 194.
45 John Murray, 'On the Total Annual Rainfall on the Land of the Globe, and the Relation of Rainfall to the Annual Discharge of Rivers', *Scottish Geographical Magazine*, III (1887), p. 65. See Meredith McKittrick, 'Theories of "Reprecipitation" and Climate Change in the Settler Colonial World', *History of Meteorology*, VIII (2017), pp. 76–7, 83–4.
46 Philip Lehmann, 'Whither Climatology? Brückner's Climate Oscillations, Data Debates, and Dynamic Climatology', *History of Meteorology*, VII

(2015), pp. 49–69; Nico Stehr and Hans von Storch, eds, *Eduard Brückner: The Sources and Consequences of Climate Change and Climate Variability in Historical Times* (Dordrecht, 2000).

47 Cited in McKittrick, 'Theories of "Reprecipitation"', p. 84. See Eduard Brückner, 'Über die Herkunft des Regens', *Verhandlungen des VII Internationalen Geographenkongresses*, vol. II (1899), pp. 412–20.

48 John Ioannidis, 'Why Most Published Research Findings Are False', *Plos Medicine*, II/8 (2005).

49 A null hypothesis was an alternative explanation, be it randomness or an unknown influence, that could be statistically defined prior to the experiment. In the tea-drinking experiment, Fisher's null hypothesis was that the lady could tell if the milk had been added before or after. One of Fisher's colleagues reportedly said that she did identify all eight cups, thus disproving the null hypothesis. See David Salsburg, *The Lady Tasting Tea: How Statistics Revolutionized Science in the Twentieth Century* (New York, 2001).

50 Phillip Lehmann, 'Average Rainfall and the Play of Colors: Colonial Experience and Global Climate Data', *Studies in History and Philosophy of Science Part A*, LXX (2018), pp. 38–49.

51 See Golinski, *British Weather and the Climate of Enlightenment*.

52 Coen, *Climate in Motion*.

53 For a clear description of the history of the network see David E. Pedgley, 'The British Rainfall Organization, 1859–1919', *Special Issue: The 150th Anniversary of the British Rainfall Organization*, LXV/5 (2010), pp. 115–17. Also see Katharine Anderson, *Predicting the Weather: Victorians and the Science of Meteorology* (Chicago, IL, and London, 2005), p. 100.

54 Pedgley, 'The British Rainfall Organization', p. 116.

55 Anderson, *Predicting the Weather*, p. 258.

56 George Symons, 'On the Climates of the Various British Colonies', *Symons Monthly Meteorological Magazine*, CXXXVI/May (1877), pp. 59–60. Cited from Anderson, *Predicting the Weather*, p. 258.

57 George Symons, *British Rainfall 1874* (London, 1875), p. 7.

58 Henry F. Blanford, *The Indian Meteorologist's Vade-Mecum Part I: Instructions to Observers* (1876), p. 42.

59 Ward, 'The Influence of Forests', p. 324.

60 Ibid., p. 323.

61 Abbe, 'Is Our Climate Changing?', p. 685.

62 Ibid.

63 Moore, *A Report on the Influence of Forests on Climate and Floods*, p. 5. Italic emphasis in original quote.

64 See Brett Bennett and Frederick Kruger, *Forestry and Water Conservation in South Africa: History, Science and Policy* (Canberra, 2015).

65 M. Hill, *Note on an Enquiry by the Government of India into the Relation between Forests and Atmospheric and Soil Moisture in India* (Calcutta, 1916).

66 Ibid., p. 4.

67 Ibid., p. 9.
68 Ibid., pp. 10–11.
69 Ibid., p. 41.
70 Ibid., p. 40.
71 Ibid., p. 38.
72 Ibid., p. 41.
73 V. K. Saberwal, 'Science and the Desiccationist Discourse of the 20th Century', *Environment and History*, IV/3 (1998), pp. 309–43.
74 James Beattie, *Empire and Environmental Anxiety: Health, Science, Art and Conservation in South Asia and Australasia, 1800–1920* (New York, 2011), p. 174.
75 Moore, *A Report on the Influence of Forests on Climate and Floods*, p. 9.
76 Ibid., p. 4.
77 'False Reasoning Injure a Cause of Conservation', *Washington Post*, 19 January 1910.
78 Dodds, 'The Stream-Flow Controversy', p. 66.
79 Gifford Pinchot, 'Government Forestry Abroad', *Publications of American Economic Association*, VI/3 (1891), p. 34.

5 Saving the World from Deserts

1 Richard St Barbe Baker, *Sahara Challenge* (London, 1954), p. 125.
2 For the term 'geographic imaginaries', see Martin Mahony and Samuel Randalls, eds, *Weather, Climate and the Geographical Imagination: Placing Atmospheric Knowledge* (Pittsburgh, PA, 2020). Diana Davis's works are the most systematic historical treatments of desertification, and they link history to science and policy today. See Diana K. Davis, *The Arid Lands: History, Power, Knowledge* (Cambridge, MA, and London, 2016); Diana K. Davis, *Resurrecting the Granary of Rome: Environmental History and French Colonial Expansion in North Africa* (Athens, OH, 2007). Other key works on deserts and arid lands include Meredith McKittrick, 'Talking about the Weather: The Language of Environmental Crisis in South Africa, 1915–1945', *Environmental History* (2018), pp. 3–27; McKittrick, 'Making Rain, Making Maps: Competing Geographies of Water and Power in 19th-Century Southwestern Africa', *Journal of African History*, LVIII/2 (2017), pp. 187–212; McKittrick, 'An Empire of Rivers: Climate Anxiety, Imperial Ambition, and the Hydropolitical Imagination in Southern Africa, 1919–1945', *Journal of Southern African Studies*, XLI/3 (2015), pp. 485–504; McKittrick, 'Theories of "Reprecipitation" and Climate Change in the Settler Colonial World', *History of Meteorology*, VIII (2017), pp. 74–94; Phillip Lehmann, 'Average Rainfall and the Play of Colors: Colonial Experience and Global Climate Data', *Studies in History and Philosophy of Science Part A*, LXX (2018), pp. 38–49; Kate Showers, *Imperial Gullies: Soil Erosion and Conservation in Lesotho* (Athens, OH, 2005); Ruth Morgan, *Running Out? Water in Western Australia* (Perth, 2015).
3 For nuclear fears see Jacob Hamblin, *Arming Mother Nature: The Birth of Catastrophic Environmentalism* (Oxford, 2013); Matthias Dörries, 'The

Politics of Atmospheric Sciences: "Nuclear Winter" and Global Climate Change', *Osiris*, XXVI (2011). For glaciers see Mark Carey, 'The History of Ice: How Glaciers Became an Endangered Species', *Environmental History*, XII (2007), pp. 497–527. For objects re-entering Earth, see Lisa Rand, 'Falling Cosmos: Nuclear Re-Entry and the Environmental History of Earth Orbit', *Environmental History*, IV/1 (2019), pp. 78–103.

4 For an overview of Lowell's ideas see Robert Crossley, 'Percival Lowell and the History of Mars', *Massachusetts Review*, XLI (2000), pp. 297–318; David Weintraub, *Life on Mars: What to Know before We Go* (Princeton, NJ, 2018), pp. 56–69.

5 Percival Lowell, *The Soul of the Far East* (Boston, MA, and New York, 1888).

6 Michael A. Amundson, 'Seeing Arizona, Imagining Mars: Deserts, Canals, Global Climate Change and the American West', *Journal of Arizona History*, LVIII/4 (2017), pp. 331–50.

7 Percival Lowell, 'Life on Mars: Address by Professor Lowell', *Daily Mail*, 31 February 1910.

8 Percival Lowell, 'Planet Disclosures by Photography: Professor Lowell on a Marvellous Feat', *Wallace Online*, https://wallace-online.org, 9 April 2010.

9 Robert Cromie, *A Plunge into Space* (London and New York, 1890), p. 87.

10 Anonymous, 'The Forests of the Planet Mars', *Indian Forester*, XXXIII (1907), p. 725. See also Gregory Barton, *Empire Forestry and the Origins of Environmentalism* (Cambridge, 2002), p. 39.

11 Alfred Russel Wallace, *Is Mars Habitable? A Critical Examination of Professor Percival Lowell's Book, 'Mars and Its Canals' with an Alternative Explanation* (London, 1907).

12 Ibid., p. 79.

13 Ibid., pp. 79–8.

14 William Sheehan and Jim Bell, *Discovering Mars: A History of Observation and Exploration of the Red Planet* (Tucson, AZ, 2021).

15 Stuart Jeffries, 'We Should Be Scared Stiff', *The Guardian*, www.theguardian.com, 16 March 2007; 'Mars Is Earth's Future', *Meadow Creek Valley*, www.meadowcreekvalley.wordpress.com, 3 October 2016; Xinhua, 'China Focus: Chinese Scientists Eye Transforming Mars after Successful Sand Control', *China Daily*, www.chinadaily.com.cn, 12 October 2018.

16 Warren E. Gates, 'The Spread of Ibn Khaldûn's Ideas on Climate and Culture', *Journal of the History of Ideas*, XXVIII/3 (1967), pp. 415–22.

17 Ben Beaumont-Thomas, 'Tataouine, Town in Tunisia That Inspired Star Wars, Becomes Isis Waypoint – Reports', *The Guardian*, www.theguardian.com, 25 March 2015.

18 FAO/UNESCO/WMO, 'World Map of Desertification', *United Nations Conference on Desertification* (Nairobi, 1977).

19 Charles Lyell, 'The Glacial Period: From an Address by Charles Lyell Delivered before the British Association', *The New York Times*, www.nytimes.com, 9 October 1864.

20 Ibid.
21 Darashaw Nosherwan Wadia, *The Post-Glacial Desiccation of Central Asia: Evolution of the Arid Zone of Asia* (Paris, 1960).
22 M. Charles Hertz, 'The Mission of Captain Roudaire in Tunisia, and the Interior Sea', *L'Explorateur*, III/59 (1876), pp. 273–6.
23 Elisee Reclus, *The Universal Geography: Earth and its Inhabitants*, vol. XI: *North-West Africa* (London, 1876), p. 78.
24 John D. Champlin Jr, 'The Proposed Inland Sea in Algeria', *Popular Science Monthly*, VIII (1876), pp. 666–7.
25 McKittrick, 'An Empire of Rivers', pp. 485–504.
26 See 'Artificial Lakes in Africa', *New York Times*, 20 June 1886.
27 George W. Plympton, 'Flooding the Sahara', *Science*, VII/176 (1886), pp. 542–4.
28 Jules Vernes, *Invasion of the Sea*, trans. Edward Baxter (Middletown, CT, 2001), p. 51.
29 Ibid., p. 47.
30 Ibid., pp. 50–51.
31 E.H.L. Schwarz, *The Kalahari; or, The Thirstland Redemption* (Cape Town, 1920), p. 4.
32 Ibid.
33 Georgina Endfield and David Nash, 'Drought, Desiccation and Discourse: Missionary Correspondence and Nineteenth-Century Climate Change in Central Southern Africa', *Geographical Journal*, CLXVIII/1 (2002), pp. 33–47.
34 William Beinart, *The Rise of Conservation in South Africa: Settlers, Livestock and the Environment, 1770–1950* (Oxford, 2003); Lance Van Sittert, '"Our Irrepressible Fellow-Colonist": The Biological Invasion of Prickly Pear (*Opuntia ficus-indica*) in the Eastern Cape, *c.* 1890–*c.* 1910', *Journal of Historical Geography*, XXVIII/3 (2002), pp. 397–410.
35 McKittrick 'Talking about the Weather'; McKittrick, 'An Empire of Rivers'; McKittrick, 'Theories of "Reprecipitation"'.
36 Beinart, *The Rise of Conservation in South Africa*, pp. 77–86, 102–11.
37 Brett Bennett and Frederick Kruger, *Forestry and Water Conservation in South Africa: History, Science and Policy* (Canberra, 2015), p. 44.
38 Wessell Visser, 'White Settlement and Irrigation Schemes: C. F. Rigg and the Founding of Bonnievale in the Breede River Valley, 1900–*c.* 1953', *New Contree*, LXVII (2013), pp. 4–5.
39 Beinart, *The Rise of Conservation in South Africa*, pp. 159–60.
40 This would be accomplished not by damming and turning the Zambezi itself, but by damming and turning a number of its tributaries, particularly the Chobe and Kunene rivers. See Schwarz, *Kalahari*, p. 122.
41 Ibid.
42 Ibid.
43 Ibid., pp. 9–10.

44 Ibid., p. 23.
45 Beinart, *The Rise of Conservation in South Africa*, pp. 249–59.
46 South Africa (Union of), *Desert Encroachment Committee Report* (Pretoria, 1951) p. 6.
47 Ibid., p. 6 and 8.
48 Ibid., p. 12.
49 Ibid., p. 17.
50 McKittrick, 'An Empire of Rivers', p. 488.
51 Davis, *The Arid Lands*, p. 122. Davis inserted the '[and Sahel]'.
52 Ibid., p. 124.
53 Edward Percy Stebbing, 'The Encroaching Sahara', *Geographical Journal*, LXXV/6 (1935), pp. 506–24.
54 Henri Duverier, *Les Touareg du Nord* (Paris, 1864).
55 Edward Percy Stebbing, 'The Threat of the Sahara', *Journal of the Royal African Society*, XXXVI/145 (1937), p. 5.
56 Edward Percy Stebbing, *The Forests of West Africa and the Sahara: A Study of Modern Conditions* (London and Edinburgh, 1937), p. 66.
57 Stebbing, 'The Encroaching Sahara', p. 21.
58 Stebbing, *The Forests of West Africa*, p. 11.
59 Ibid., p. 11.
60 Edward Percy Stebbing, 'Africa and Its Intermittent Rainfall: The Role of the Savannah Forest', *Journal of the Royal African Society*, XXXVII/149 (1938), p. 3.
61 Ibid., p. 23.
62 Brett M. Bennett and Gregory A. Barton, 'E.H.F. Swain's Vision of Forest Modernity', *Intellectual History Review*, XXI/2 (2011), pp. 147–9.
63 Baker, *Sahara Challenge*, p. 13.
64 Ibid., p. 14.
65 Ibid., pp. 27–8.
66 Ibid., p. 18.
67 Ibid., p. 23.
68 Ibid., p. 24.
69 Ibid., pp. 27–8.
70 Ibid., p. 70.
71 Ibid., pp. 76–7.
72 Ibid., p. 111.
73 Ibid.
74 Ibid., p. 113.
75 Ibid., p. 124.
76 Richard St Barbe Baker, *Kamiti* (New York, 1958), p. 4.
77 Ibid., p. 25.
78 Ibid., p. 37.
79 Ibid., pp. 47, 49 and 53.
80 Ibid., p. 58.
81 Ibid., p. 70.

82 Ibid., p. 62.
83 Ibid., p. 63.
84 Ibid., p. 70.
85 Ibid., p. 115.

6 How Dreams of Reclaiming Deserts Evaporated

1 Stephen Cattle, 'The Case for a South Eastern Australia Dust Bowl, 1893–45', *Aeolian Research*, XXI (2016), pp. 1–20.
2 James Fleming, *Fixing the Sky: The Checkered History of Weather and Climate Control* (New York, 2012), pp. 77–109.
3 Ibid., pp. 132–3.
4 J. C. Jensen, 'The Relation between Surface Evaporation from Lakes and Ponds to Precipitation from Local Thunderstorms in the Drought Area', *Bulletin of the American Meteorological Society*, XVI (1935), pp. 142–5 (p. 143).
5 Ibid., p. 145.
6 E. T. Quayle, 'Possibilities of Modifying Climate by Human Agency', *Proceedings of the Royal Society of Victoria*, XXXIII (1921).
7 The classic work on this subject is Donald Worster, *Dust Bowl: The Southern Plains in the 1930s* (New York, 1979).
8 Section 1, Timber Culture Act, 3 March 1873. See *The United States Statutes at Large*, XVII, Forty-Second Congress Session II. Ch. 274–7 (Boston, MA, 1873), p. 605.
9 Frederic E. Clements and Ralph W. Chaney, *Environment and Life in the Great Plains* (Washington, DC, 1936), p. 41.
10 Myron P. Gutmann, William J. Parton, Geoff Cunfer and Ingrid C. Burke, 'Population and Environment in the U.S. Great Plains', *National Research Council (U.S.) Panel on New Research on Population and the Environment*, V (2005).
11 Frederic E. Clements, 'Rainfall and Climatic Cycles', in *Carnegie Institute of Washington: Year Book No. 22* (Washington, DC, 1923), pp. 318–19.
12 Benjamin Holzman, 'Sources of Moisture for Precipitation in the United States', *United States Department of Agriculture* (1937), p. 15.
13 An imperial committee came to this conclusion. See Department of Agriculture and Forestry, 'Forests in Relation to Climate, Water Conservation and Erosion' (Pretoria, 1935), pp. 8–9.
14 National Resources Board, 'A report on national planning and public works in relation to natural resources and including land use and water resources with findings and recommendations', 1 December 1934. Submitted to the President in accordance with Executive order no. 6777, 30 June 1934 (Washington, DC). Both quotes come from Benton and also Holzman.
15 Stephen Legg, 'Debating the Climatological Role of Forests in Australia, 1827–1949', in *Climate, Science and Colonization: Histories from Australia and New Zealand*, ed. James Beattie, Matthew Henry and Emily O'Gorman (New York, 2014), pp. 119–36.

16 C. W. Thornthwaite and Benjamin Holzman, 'Measurement of Evaporation from Land and Water Surfaces', *United States Department of Agriculture* (1942), p. 64.
17 Susan Broomhall, 'Feeling Divine Nature: Natural History, Emotion and Bernard Palissy's Knowledge Practice', in *Natural History in Early Modern France: The Poetics of an Epistemic Genre*, ed. Raphaële Garrod and Paul J. Smith (Leiden, 2018), p. 60.
18 Jamie Linton, *What Is Water? The History of a Modern Abstraction* (Vancouver, 2010), p.115.
19 Wilfried Brutsaert, *Hydrology: An Introduction* (Cambridge, 2005).
20 Ibid.
21 Edmund Halley, 'An Account of the Evaporation of Water, as It Was Experimented in Gresham College in the Year 1693. With some observations Thereon', *Phil Trans*, XVIII (1693), pp. 181–90.
22 George Marsh, *Man and Nature: Physical Geography as Modified by Human Action* (London, 1864), p. 183.
23 Ibid.
24 Baldwin Latham, 'Upon the Supply of Water to Towns', *Civil Engineer and Architects Journal*, XXVIII (1865), pp. 20–24.
25 Editor, 'The Evaporation of Ocean Water', *Popular Mechanics* (1913), p. 724.
26 George S. Benton, Robert T. Blackburn and Vernon O. Snead, 'The Role of the Atmosphere in the Hydrologic Cycle', *Weatherwise*, II/2 (1949), pp. 99–103 (p. 101).
27 Jamie Linton, 'Is the Hydrologic Cycle Sustainable? A Historical–Geographical Critique of a Modern Concept', *Annals of the Association of American Geographies*, XCVIII/3 (2008), p. 636.
28 For 33 per cent, see Robert E. Horton, 'Hydrologic Interrelations between Lands and Oceans', *Eos Trans, AGU*, XXIV (1943), p. 764. For 75 per cent see Benjamin Holzman, *Sources of Moisture for Precipitation in the United States* (Washington, DC, 1937), pp. 2, 9.
29 For one-third see T. Vaville, 'Basic Principles of Water Behavior', in 'Headwaters: Control and Use', a paper presented at the Upstream Engineering Conference, Washington, DC, 22 and 23 September 1936, pp. 1–10. For two-thirds, see C.E.P. Brooks, 'The Influence of Forests on Rainfall and Run-off', *Royal Meteorological Society Quarterly Journal*, LIV/225 (1928), pp. 1–17. All discussed originally by Holzman, *Sources of Moisture for Precipitation*, p. 11.
30 Cited from Linton, *What Is Water?*, p. 136.
31 Significant efforts went into measuring the 'water balance', a formula that measured the total input (precipitation) and outflow (evaporation, transpiration, stream flow out of catchment) of a catchment or region. The water balance can be expressed somewhat simply, but in practice measuring water accurately and finding the mathematical equations to describe real world situations has proved extremely difficult because measuring all water

across a landscape is difficult, if not technically impossible given the scale and difficulty of measuring water.

32 Robert E. Horton, 'A New Evaporation Formula Developed: Empirical Statement Based on Physical Laws Agrees with Observed Facts and Is Held to Be an Improvement over Existing Formulas', *Engineering News Record*, LXXVII (1917), p. 196.

33 Robert Horton, 'Ebermayer's Experiments on Forest Meteorology', *Michigan Engineering Society* (1911), pp. 65–6, 66–89.

34 Ibid.

35 Douglas Golding, 'The Effect of Forests on Precipitation', *Forestry Chronicle*, XLVI/5 (1970), p. 399.

36 Vazken Andréassian, 'Waters and Forests: From Historical Controversy to Scientific Debate', *Journal of Hydrology*, XXXI/3 (1933), pp. 296–307 (p. 296).

37 W. C. Lowdermilk, 'Forests and Streamflow: A Discussion of the Hoyt–Troxell Report', *Journal of Forestry*, XXXI/3 (1933), pp. 296–307 (p. 296). Cited in Brett Bennett and Frederick Kruger, *Forestry and Water Conservation in South Africa: History, Science and Policy* (Canberra, 2015), p. 188.

38 CCC, *Woodsmanship for the Civilian Conservation Corps* (1941), p. 12.

39 Cited in Brett Bennett and Frederick Kruger, 'Ecology, Forestry and the Debate over Exotic Trees in South Africa', *Journal of Historical Geography*, XLII (2013), p. 107. Originally cited in Anon, 'Forest Planting in South Africa: Colonel Reitz's Alarm at the Effect of Erosion', *Natal Mercury*, 3 September 1935.

40 Cited in Bennett and Kruger, 'Ecology, Forestry and the Debate over Exotic Trees in South Africa', p. 107. Anon, 'Forestry Research Essential: General Smuts Stresses Need for Scientific Development', *Rand Daily Mail*, 9 September 1935.

41 Department of Agriculture and Forestry, *Forests in Relation to Climate, Water Conservation and Erosion* (Pretoria, 1935), p. 9.

42 Bennett and Kruger, *Forestry and Water Conservation in South Africa*, p. 224.

43 Murray C. Peel, Thomas McMahon and Brian L. Finlayson, 'Vegetation Impact on Mean Annual Evapotranspiration at a Global Catchment Scale', *Water Resources Research*, XLVI (2010), pp. 1–16.

44 Alden R. Hibbert, 'Forest Treatment Effects on Water Yield', *International Symposium on Forest Hydrology: Proceedings of a National Science Foundation on Advanced Science Seminar Held at the Pennsylvania State University, Pennsylvania, August 29–September 10, 1965* (Oxford and New York, 1966). The article was republished in *United States Congress. Senate. Committee on Interior and Insular Affairs Hearing 1971*, vol. II (Washington, DC, 1971), p. 887.

45 Ibid.

46 Ibid.

47 This was supported by further hydrological theoretical work. H. L. Penman, *Vegetation and Hydrology. Commonwealth Agricultural Bureaux Technical*

Communications No. 53, Commonwealth Bureau of Soils, Harpenden (London, 1963).

48 For a detailed analysis of scientific theories of precipitation prior to the 1900s see William Edward Knowles Middleton, 'A History of the Theories of Rain and Other Forms of Precipitation', *Quarterly Journal of the Royal Meteorological Society*, XCII/394 (1966), p. 206.

49 Michael J. Doherty, 'James Glaisher's 1862 Account of Balloon Sickness: Altitude, Decompression Injury, and Hypoxemia', *Neurology*, XXV (2003), pp. 1016–80.

50 Peter Moore, *The Weather Experiment: The Pioneers Who Sought to See the Future* (London, 2015), p. 266.

51 Ibid.

52 Frederik Nebeker, *Calculating the Weather: Meteorology in the 20th Century*, 1st edn (San Diego, CA, 1995), p. 48.

53 Kristine Harper, *Weather by the Numbers: The Genesis of Modern Meteorology* (Cambridge, MA, and London, 2008), pp. 51–61.

54 Ibid., p. 2.

55 Ibid., p. 63.

56 Vilhelm Bjerknes, 'The Structure of the Atmosphere When Rain Is Falling', *Quarterly Journal of the Royal Meteorological Society*, XLVI/194 (1920), p. 119.

57 Ibid., pp. 119–40.

58 Ibid.

59 Alexander McAdie, *Making the Weather* (New York, 1923), p. 35.

60 Robert Marc Friedman, *Appropriating the Weather: Vilhelm Bjerknes and the Construction of a Modern Meteorology* (Ithaca, NY, and London, 1989), p. 1.

61 Holzman, 'Sources of Moisture for Precipitation', p. 3.

62 Ibid., p. 1.

63 James R. Fleming, 'Sverre Petterssen, the Bergen School and Forecasts of D-Day', *Proceedings of the International Commission on the History of Meteorology*, I/1 (2004), pp. 1–9.

64 Georg Wüst, 'Verdunstung und Niederschlag auf der Erde', *Zeitschrit der Gesellschaft für Erdkunde zu Berlin* (Berlin, 1922), pp. 35–43.

65 Holzman, 'Sources of Moisture for Precipitation', p. 2.

66 Ibid.

67 Ibid.

68 Ibid., p. 3.

69 Ibid.

70 Ibid.

71 Ibid., p. 2.

72 Ibid., pp. 5–6.

73 Ibid., p. 38.

74 Ibid., p. 12.

75 Ibid.

76 Ibid., p. 14.

77 George S. Benton, 'The Role of the Atmosphere in the Hydrologic Cycle', *Weatherwise*, II/2 (1949), pp. 99–103 (p. 101).
78 'Oral History Interview with George S. Benton', *OpenSky*, www.opensky.ucar.edu, 27 May 1991.
79 Ibid.
80 Benton, 'The Role of the Atmosphere in the Hydrologic Cycle', pp. 99–103 (p. 101).
81 D. M. Schultz and R. M. Friedman, 'Bergoron, Tor Harold Percival', *New Dictionary of Scientific Biography* (New York, 2007), pp. 245–50 (p. 248).
82 Roscoe R. Braham Jr, 'Formation of Rain: A Historical Perspective', in James Roger Fleming, *Historical Essays on Meteorology, 1919–1995* (Boston, MA, 1996), pp. 192–4.
83 Shultz and Friedman, 'Bergoron, Tor Harold Percival', pp. 248–9.
84 Jacob Hamblin, *Arming Mother Nature: The Birth of Catastrophic Environmentalism* (Oxford, 2013).
85 Kristine C. Harper and Ronald E. Doel, 'Environmental Diplomacy in the Cold War', in *Environmental Histories of the Cold War*, ed. J. R. McNeill and Corinna Unger (Cambridge, 2010), pp. 129–30.
86 George S. Benton, Robert T. Blackburn and Vernon O. Snead, 'The Role of the Atmosphere in the Hydrological Cycle', *Trans. Am. Geoph. Union*, XXXI (1950), pp. 61–73.
87 George Benton and Robert Blackburn, 'A Comparison of Precipitation from Maritime and Continental Air', *Bulletin of the American Meteorological Society*, XXXI/7 (1950), pp. 254–6.
88 Congressional Record, 98 (21 June 1952), p. 7777. Cited from Jedediah S. Rogers, 'Project Skywalker', *Bureau of Reclamation* (2009), p. 9.
89 Colin Ward, 'Cloud-Seeding', *CSIRO*, www.csiropedia.csiro.au, accessed 8 December 2021.
90 Engineering and Research Centre, Division of Atmospheric Water Resource Management, Project Skywater: Environmental Impact Statement, I (1977), p. 46.
91 Kristine Harper, *Make It Rain: State Control of the Atmosphere in Twentieth-Century America* (Chicago, IL, 2017), p. 149.
92 Stephen Cole and Roscoe R. Braham, 'Tape Recorded Interview Project: Interview of Roscoe R. Braham', *American Meteorological Society University Corporation for Atmospheric Research* (2002), p. 4.
93 Vincent Schaefer, 'After a Quarter Century', *Weather Modification Association*, I/1 (1969), pp. 1–4 (p. 1).
94 Ibid., p. 3.
95 Ibid., pp. 26–7.
96 James McDonald, 'Evaluation of Weather Modification Field Tests', in *Weather Modification, Science and Public Policy*, ed. R. G. Fleagle (Seattle, WA, 1969), pp. 44–5.
97 Ibid.

98 Harper, *Make It Rain*, p. 7.
99 Fleming, *Fixing the Sky*, p. 9.
100 James McDonald, 'The Evaporation-Precipitation Fallacy', *Weather*, XVII/5 (1962), pp. 168–77 (p. 168).
101 Also see his letter, James McDonald, 'The Evaporation-Precipitation Fallacy', *Weather*, 17 (1962), p. 216.
102 Cairo Correspondent, 'Lake in the Desert', *The Economist*, 3 March 1965, p. 27.
103 McDonald, 'The Evaporation-Precipitation Fallacy', p. 169.
104 Ibid., p. 176.
105 Charles Pereira, *Land Use and Water Resources* (Cambridge, 1973), p. 5.
106 Golding, 'The Effect of Forests on Precipitation', p. 401.

7 The Revival of the Forest–Rainfall Connection

1 Benjamin W. Abbot et al., 'Human Domination of the Global Water Cycle Absent from Depictions and Perceptions', *Nature Geoscience*, XII (2019), pp. 533–40.
2 Paul N. Edwards, *A Vast Machine: Computer Models, Climate Data and the Warming of Global Politics* (Cambridge, MA, 2013), p. xvi.
3 David Ellison et al., 'Trees, Forests and Water', *Global Environmental Change*, XLIII (2017), pp. 51–61 (p. 51).
4 A. M. Makarieva, V. G. Gorshkov and Bai-Lan Li, 'Conservation of Water Cycle on Land via Restoration of Natural Closed-Canopy Forests: Implications for Regional Landscape Planning', *Ecological Research*, XXI/6 (2006), pp. 897–906; A. M. Makarieva and V. G. Gorshkov, 'Biotic Pump of Atmospheric Moisture as Driver of the Hydrological Cycle on Land', *Hydrology and Earth System Sciences*, XI/2 (2007), pp. 1013–33; Douglas Sheil and Daniel Murdiyarso, 'How Forests Attract Rain: An Examination of a New Hypothesis', *Bioscience*, LIX/4 (2009), pp. 341–7; David Ellison, Martyn N. Futter and Kevin Bishop, 'On the Forest Cover–Water Yield Debate: From Demand to Supply Side Thinking', *Global Change of Biology*, XIII/3 (2011), pp. 806–20; Douglas Sheil, 'How Plants Water Our Planet: Advances and Imperatives', *Trends in Plant Science*, XIX/4 (2014), pp. 209–11.
5 Ellison, Futter and Bishop, 'On the Forest Cover', pp. 806–20.
6 Also see the water footprint literature for similar concepts. See Rick Hogeboom, 'The Water Footprint Concept and Water's Grand Environmental Challenges', *One Earth*, II/3 (2020), pp. 218–22.
7 J. M. Bosch and J. D. Hewlett, 'A Review of Catchment Experiments to Determine the Effect of Vegetation Changes on Water Yield and Evapotranspiration', *Journal of Hydrology*, LV/1–4 (1982), pp. 3–23; Vazken Andreassian, 'Water and Forests: From Historical Controversy to Scientific Debate', *Journal of Hydrology*, CCXCI/1–2 (2004), pp. 1–27.
8 Ellison, Futter and Bishop, 'On the Forest Cover', pp. 806–20.
9 For an interesting discussion on this topic see David Ellison et al., 'Europe's Forest Sink Obsession', EGU General Assembly 2022, Vienna, Austria,

23–27 May 2022, EGU22-8950 (2020), https://doi.org/10.5194/egusphere-egu22-8950, pp. 1–14.

10 See Brett Bennett, *Plantations and Protected Areas: A Global History of Forest Management* (Cambridge, MA, 2015).

11 Iain McIntyre, 'From the Local to the Global and Back Again: The Rainforest Information Centre and Transnational Environmental Activism in the 1980s', in *The Transnational Activist: Transformations and Comparisons from the Anglo-World since the Nineteenth Century,* ed. Stefan Berger and Sean Scalmer (Cham, 2017). For a somewhat longer history, see Seth Garfield, 'The Brazilian Amazon and the Transnational Environment, 1940–1990', in *Nation-States and the Global Environment: New Approaches to International Environmental History*, ed. Erika Marie Bsumek, David Kinkela and Mark Atwood Lawrence (New York, 2013), pp. 228–51.

12 See www.merriam-webster.com, accessed 1 February 2022.

13 David Arnold, 'The Place of the "Tropics" in Western Medical Ideas since 1750', *Tropical Medicine and International Health*, II/4 (2007), pp. 303–13.

14 See Richard Grove, *Green Imperialism: Colonial Expansion, Tropical Island Edens and the Origins of Environmentalism, 1600–1860* (Cambridge, 1995).

15 George Marsh, *Man and Nature: Physical Geography as Modified by Human Action* (London, 1864), p. 184.

16 Ibid., p. 160.

17 Julius von Hann, *Handbook of Climatology*, vol. I (London, 1903), p. 194.

18 W. G. Kendrew, *Climatology: Treated Mainly in Relation to Distribution in Time and Place*, 3rd edn (Oxford, 1942), p. 194.

19 'Grow Trees for Rain', *The Straits Times* (Singapore), 21 August 1963, p. 4.

20 'They Bring Rain Too, Says Minister Wok', *Straits Times* (Singapore), 3 February 1964, p. 11.

21 Peter Dauvergne, *Historical Dictionary of Environmentalism,* 2nd edn (Lanham, MD, 2016).

22 Ricardo Adrogué, Martin Cerisola and Gaston Gelos, *Brazil's Long-Term Growth Performance: Trying to Explain the Puzzle*, IMF Working Paper No. 2006/282 (2006), www.imf.org/en, accessed 10 October 2023.

23 Felipe Fernandes Cruz, 'Napalm Colonization: Native Peoples in Brazil's Aeronautical Frontiers', *Hispanic American Historical Review*, CI/3 (2021), pp. 461–89; Eve E. Buckley, *Technocrats and the Politics of Drought and Development in Twentieth-Century Brazil* (Chapel Hill, NC, 2017).

24 M.A.M. Dantes et al., 'Sciences in Brazil: An Overview from 1870–1920', in *Brazilian Studies in Philosophy and History of Science: An Account of Recent Works*, ed. Antonio Videira and Décio Krause (Dordrecht, 2001), p. 100.

25 Thomas Rudel, *Tropical Forests: Regional Paths of Destruction and Regeneration in the Late Twentieth Century* (New York, 2005), pp. 51–8.

26 Michael Williams, *Deforesting the Earth* (Chicago, IL, 2003), pp. 465–82.

27 Reginald E. Newell et al., *The General Circulation of the Tropical Atmosphere and Interactions with Extratropical Latitudes*, vol. I (Cambridge, MA, and London, 1973).
28 Luiz Carlos Baldicero Molion, 'A Climatonomic Study of the Energy and Moisture Fluxes of the Amazonas Basin with Consideration to Deforestation Effects', PhD thesis, University of Wisconsin-Madison (1975), p. 10.
29 José Marques et al., 'Precipitable Water and Water Vapor Flux between Belém and Manaus', *Acta Amazonica*, VII/3 (1977), pp. 361–2.
30 Irving Friedman, 'The Amazon Basin, Another Sahel?', *Science*, CXCVII/4298 (1977), p. 7.
31 For instance, oxygen isotopes usually have an atomic mass of 16 (8 protons and 8 neutrons) but there are some with 18 (8 protons and 10 neutrons) and an even rarer 17 (8 protons, 9 neutrons).
32 Samuel Epstein and Toshiko Mayeda, 'Variation of O^{18} Content of Waters from Natural Sources', *Geochimica et Cosmochimica Acta*, IV/5 (1973), pp. 213–24.
33 Harmon Craig, 'Isotopic Variations in Meteoric Waters', *Science,* CXXXIII/3465 (1961), pp. 1702–3.
34 W. Dansgaard, 'Stable Isotopes in Precipitation', *Tellus*, XVI/4 (1964), pp. 436–68.
35 Eneas Salati et al., 'Recycling of Water in the Amazon Basin: An Isotopic Study', *Water Resources Research*, XV/5 (1979), pp. 1250–58.
36 Ibid., p. 1258.
37 Eneas Salati and Peter B. Vose, 'Amazon Basin: A System in Equilibrium', *Science*, CCXXV/4658 (1984), pp. 130–33.
38 Ibid., p. 137.
39 Eneas Salati, Thomas Lovejoy and Peter. B. Vose, 'Precipitation and Water Recycling in Tropical Rain Forests with Special Reference to the Amazon Basin', *Commission on Ecology Occasional Paper* (1983), p. 67.
40 Ibid., p. 361.
41 Robert White, 'Climate at the Millennium', in *Proceedings of the World Climate Conference: A Conference of Experts on Climate and Mankind* (Geneva, 1979), pp. 1–14 (p. 4).
42 Juan Burgos, 'Renewable Resources and Agriculture in Latin America in Relation to the Stability of Climate', in *Proceedings of the World Climate Conference: A Conference of Experts on Climate and Mankind*, p. 536.
43 Ann Henderson-Sellers and Vivien Gornitz, 'Possible Climatic Impact of Land Cover Transformations, with Particular Emphasis on Tropical Deforestation', *Climatic Change*, VI (1984), pp. 231–57.
44 Diana K. Davis, *The Arid Lands: History, Power, Knowledge* (Cambridge, MA, and London, 2016).
45 Jule Gregory Charney, 'Dynamics of Deserts and Droughts in the Sahel', *Quarterly Journal of the Royal Meteorological Society*, CI/428 (1975),

pp. 193–202; Jule Charney et al., 'A Comparative Study of the Effects of Albedo Change on Drought in Semi-Arid Regions', *Journal of Atmospheric Science*, XXXIV/9 (1977), pp. 1366–85.
46 Gerald L. Potter et al., 'Possible Climatic Impact of Tropical Deforestation', *Nature*, XCCLVIII (1975), pp. 697–8.
47 Carl Sagan, Owen B. Toon and James B. Pollack, 'Anthropogenic Albedo Changes and the Earth's Climate', *New Series*, CCVI/4425 (1979), pp. 1363–68 (p. 1367).
48 Ibid.
49 Potter et al., 'Possible Climatic Impact of Tropical Deforestation', pp. 697–8.
50 Henderson-Sellers and Gornitz, 'Possible Climatic Impact of Land Cover Transformations', pp. 231–57; Robert E. Dickenson and Ann Henderson-Sellers, 'Modelling Tropical Deforestation: A Study of GCM Land-Surface Parametrizations', *Quarterly Journal of the Royal Meteorological Society*, CXIV/480 (1988), pp. 439–62.
51 W. James Shuttleworth, 'Evapotranspiration from Amazonian Rainforests', *Proceedings of the Royal Society of Biological Sciences*', CCXXXIII (1988), pp. 321–46; John H. C. Gash and W. James Shuttleworth, 'Tropical Deforestation: Albedo and the Surface Energy Rebalance', *Climatic Change*, XIX/1–2 (1991), pp. 123–33.
52 Gordan B. Bonan, 'Forests and Climate Change: Forcings, Feedbacks, and the Climate Benefits of Forests', *Science*, CCCXX/320 (2008), pp. 1444–9.
53 There still are researchers working on this question, often from a biogeophysical effects (BGP) perspective. See Michael G. Windisch, Edouard L. Davin and Sonia I. Seneviratne, 'Prioritizing Forestation Based on Biogeochemical and Local Biogeophysical Impacts', *Nature Climate Change*, XI (2021), pp. 867–71; Deborah Lawrence et al., 'The Unseen Effects of Deforestation: Biophysical Effects on Climate', *Frontiers in Forests and Global Change*, V (2022), pp. 1–13.
54 Sharon E. Nicholson, *Dryland Climatology* (Cambridge, 2011), p. 105.
55 Edouard L. Davin and Nathalie de Noblet-Ducoudré, 'Climate Impact of Global-Scale Deforestation: Radiative versus Nonradiative Process', *Journal of Climate*, XXIII/1 (2010), pp. 97–112.
56 William J. Bond, 'Ancient Grasslands at Risk', *Science*, CCCLI/6269 (2016), pp. 120–22.
57 Bonan, 'Forests and Climate Change', pp. 1444–9; Tong Jiao et al., 'Global Climate Forcing from Albedo Change Caused by Large Scale Deforestation and Reforestation: Quantification and Attribution of Geographic Variation', *Climatic Changes*, MMXVII (2017), pp. 463–76.
58 Lawrence et al., 'The Unseen Effects'.
59 Mike Hulme, *Weathered: Cultures of Climate Change* (London, 2017).
60 Guido van der Werf et al., 'CO_2 Emissions from Forest Loss', *Nature Geosciences*, II (2009), pp. 737–8. With a correction of 12 per cent from an original estimate of 20 per cent. Peter Reich, 'Biogeochemistry: Taking Stock

of Forest Carbon', *Nature Climate Changes*, I (2011), pp. 346–7; Valentin Bellassen and Sebastian Luyssaert, 'Carbon Sequestration: Managing Forests in Uncertain Times', *Nature*, DVI (2014), pp. 153–5.

61 Ellison et al., 'Trees, Forests and Water', p. 51.

62 Ruud J. van der Ent et al., 'The Importance of Proper Hydrology in the Forest Cover–Water Yield Debate: Commentary on Ellison et al.', *Global Change Biology*, XVIII/9 (2012), pp. 2677–80.

63 Michael Sanderson et al., 'Influence of EU Forests on Weather Patterns: Final Report', *Meteorological Office Devon* (2013), pp. 1–32.

64 David Ellison et al., 'Governance Options for Addressing Changing Forest-Water Relations', in *Forest and Water on a Changing Planet: Vulnerability, Adaptation and Governance Opportunities. A Global Assessment Report*, ed. Irena F. Creed and Meine van Noordwijk (Vienna, 2018), pp. 147–69 (pp. 148–51).

65 Brett Bennett and Frederick Kruger, *Forestry and Water Conservation in South Africa: History, Science and Policy* (Canberra, 2015).

66 Ruud J. van der Ent et al., 'Origin and Fate of Atmospheric Moisture over Continents', *Water Resources Research*, XLVI/9 (2010), pp. 1–12.

67 See Anne J. Hoek van Dijke et al., 'Shifts in Regional Water Availability Due to Global Tree Restoration', *Nature Geosciences*, XV (2022), pp. 363–8.

68 Supantha Paul et al., 'Weakening of Indian Summer Monsoon Rainfall Due to Changes in Land Use Land Cover', *Scientific Reports*, VI (2016), pp. 1–10.

69 Chris Smith, Jessica C. A. Baker and Dominick Spracklen, 'Tropical Deforestation Causes Large Reductions in Observed Precipitation', *Nature*, DCXV (2023), pp. 270–75.

70 Akash Ganguly et al., 'Extreme Local Recycling of Moisture via Wetlands and Forests in North-East Indian Subcontinent: A Mini-Amazon', *Scientific Reports*, XIII (2023), pp. 1–10.

71 Solomon Gebreyohannis Gebrehiwot et al., 'The Nile Basin Waters and the West African Rainforest: Rethinking the Boundaries', *Wires Water*, VI/1 (2018), pp. 1–8.

72 Hoong Chen Teo et al., 'Large-Scale Reforestation Can Increase Water Yield and Reduce Drought Risk for Water-Insecure Regions in the Asia-Pacific', *Global Change Biology*, XXVIII/21 (2022), pp. 6385–403.

73 Jean-François Bastin et al., 'The Global Tree Restoration Potential', *Science*, CCCLXV/6448 (2019), pp. 76–9.

74 Ellison, Futter and Bishop, 'On the Forest Cover'; Ellison et al., 'Trees, Forests and Water'.

75 Natasha Gilbert, 'Green Space: A Natural High', *Nature*, DXXXI (2016), pp. 56–7.

76 There are numerous studies for cities around the world. For an accessible introduction to the topic see Eric Thompson, Mitch Herian and David Rosenbaum, *The Economic Footprint and Quality-of-Life Benefits of Urban Forestry in the United States* (Lincoln, NE, 2021).

77 Robin. L. Chazdon, 'Beyond Deforestation: Restoring Forests and Ecosystem Services on Degraded Lands', *Science*, CCCXX (2008), pp. 1458–60.

78 Feng Wang et al., 'Vegetation Restoration in Northern China: A Contrasted Picture', *Land Degradation and Development*, XXXI/6 (2019), pp. 669–76; Antje Ahrends et al., 'China's Fight to Halt Tree Cover Loss', *Proceedings of the Royal Society: Biological Sciences,* CCCLXXXIV (2017), pp. 1–10.
79 Shuoyu Chen et al., 'Quantifying the Impact of Large-Scale Afforestation on the Atmospheric Water Cycle during Rainy Season over the Chinese Loess Plateau', *Journal of Hydrology*, DCXIX (2023), article 129326.
80 Ian R. Calder, 'Forests and Hydrological Services: Reconciling Public and Science Perceptions', *Land Use and Water Resources Research*, II (2002), pp. 1–12.
81 Terry Purcell, Erminielda Peron and Rita Berto, 'Why Do Preferences Differ between Scene Types?', *Environment and Behaviour*, XXXIII/1 (2001), pp. 93–106; Agnes E. van den Berg, Terry Hartig and Henk Staats, 'Preference for Nature in Urbanized Societies: Stress, Restoration and the Pursuit of Sustainability', LXIII/1 (2007), pp. 79–96; Giuseppe Carrus et al., 'Go Greener, Feel Better? The Positive Effects of Biodiversity on the Well-Being of Individuals Visiting Urban and Peri-Urban Green Areas', *Landscape and Urban Planning*, CXXXIV (2015), pp. 221–8.
82 Bond, 'Ancient Grasslands at Risk'.
83 William Bond and Guy Midgley, 'Carbon Dioxide and the Uneasy Interactions of Trees and Savannah Grasses', *Philosophical Transactions of the Royal Society of London B: Biological Sciences*, CCCLXVII/1588 (2012), pp. 601–12.
84 Bond, 'Ancient Grasslands at Risk'.
85 William J. Bond and Catherine L. Parr, Beyond the Forest Edge: Ecology, Diversity and Conservation of the Grassy Biomes', *Biological Conservation*, CXLIII/10 (2010), pp. 2395–404.
86 Arun Agrawal, Daniel Nepstad and Ashwini Chhatre, 'Reducing Emissions from Deforestation and Forest Degradation', *Annual Review of Environmental Resources*, XXXVI (2011), pp. 373–96.

Conclusion: Interpreting the History of Climate Change

1 See Chapter Two and also Gregory Cushman, 'Humboldtian Science, Creole Meteorology, and the Discovery of Human-Caused Climate Change in Northern South America', *Osiris*, XXVI (2011), pp. 19–44.
2 Ruud J. van der Ent et al., 'Origin and Fate of Atmospheric Moisture over Continents', *Water Resources Research*, XLVI/9 (2010), pp. 1–12 (p. 1).
3 Ibid., p. 8.
4 Ibid., pp. 7–8.
5 Ibid., p. 8.
6 Cushman, "Humboldtian Science', p. 43.
7 David Ellison and Chinwe Ifejika Speranza, 'From Blue to Green Water and Back Again: Promoting Tree Shrub and Forest-Based Landscape Resilience in the Sahel', *Science of the Total Environment*, DCCXXXIX (2020), pp. 1–14.

BIBLIOGRAPHY

Abbot, Benjamin W., et al., 'Human Domination of the Global Water Cycle Absent from Depictions and Perceptions', *Nature Geoscience*, XII (2019), pp. 533–40

Agrawal, Arun, Daniel Nepstad and Ashwini Chhatre, 'Reducing Emissions from Deforestation and Forest Degradation', *Annual Review of Environmental Resources*, XXXVI (2011), pp. 373–96

Ahrends, Antje, et al., 'China's Fight to Halt Tree Cover Loss', *Proceedings of the Royal Society: Biological Sciences*, CCLXXXIV (2017), pp. 1–10

Albritton Jonsson, Fredrik, *Enlightenment's Frontier: The Scottish Highlands and the Origins of Environmentalism* (New Haven, CT, 2013)

—, 'Rival Ecologies of Global Commerce: Adam Smith and the Natural Historians', *American Historical Review*, CXV (2010), pp. 1342–63

Amundson, Michael A., 'Seeing Arizona, Imagining Mars: Deserts, Canals, Global Climate Change and the American West', *Journal of Arizona History*, LVIII/4 (2017), pp. 331–50

Anderson, Katharine, *Predicting the Weather: Victorians and the Science of Meteorology* (Chicago, IL, and London, 2005)

Andreassian, Vazken, 'Water and Forests: From Historical Controversy to Scientific Debate', *Journal of Hydrology*, CCXCI/1–2 (2004), pp. 1–27

Arnold, David, 'The Place of the "Tropics" in Western Medical Ideas since 1750', *Tropical Medicine and International Health*, II/4 (2007), pp. 303–13

Baker, Richard St Barbe, *Sahara Challenge* (London, 1954)

Baldicro, Molin Luiz Carlo, 'A Climatonomic Study of the Energy and Moisture Fluxes of the Amazonas Basin with Considerations of Deforestation Effects', *Environmental Science* (1975), pp. 1–15

Barnett, Lydia, *After the Flood: Imagining the Global Environment in Early Modern Europe* (Baltimore, MD, 2019)

Barrow, Mark V., Jr, *Nature's Ghosts: Confronting Extinction from the Age of Jefferson to the Age of Ecology* (Chicago, IL, 2009)

Barton, Gregory, *Empire Forestry and the Origins of Environmentalism* (Cambridge, 2002)

—, and Brett M. Bennett, 'A Case Study in the Environmental History of Gentlemanly Capitalism: The Battle between Gentlemen Teak Merchants and State Foresters in Burma and Siam, 1827–1901', in *Africa, Empire, and Globalization: Essays in Honor of A. G. Hopkins*, ed. Toyin Falola and Emily Brownell (Durham, NC, 2011)

—, 'Forestry as Foreign Policy: Anglo-Siamese Relations and the Origins of Britain's Informal Empire in the Teak Forests of Northern Siam, 1883–1925', *Itinerario*, XXXIV/2 (2010), pp. 65–86

Bastin, Jean-François, et al., 'The Global Tree Restoration Potential', *Science*, CCCLXV/6448 (2019), pp. 76–9

Beattie, James, *Empire and Environmental Anxiety: Health, Science, Art and Conservation in South Asia and Australasia, 1800–1920* (New York, 2011)

Behringer, Wolfgang, *A Cultural History of Climate* (London, 2010)

Beinart, William, *The Rise of Conservation in South Africa: Settlers, Livestock and the Environment, 1770–1950* (Oxford, 2003)

Bellassen, Valentin, and Sebastian Luyssaert, 'Carbon Sequestration: Managing Forests in Uncertain Times', *Nature*, DVI (2014), pp. 153–5

Bennett, Brett, *Plantations and Protected Areas: A Global History of Forest Management* (Cambridge, MA, 2015)

—, and Frederick Kruger, 'Ecology, Forestry and the Debate over Exotic Trees in South Africa', *Journal of Historical Geography*, XLII (2013), pp. 100–109

—, *Forestry and Water Conservation in South Africa: History, Science and Policy* (Canberra, 2015)

Bonan, Gordan B., 'Forests and Climate Change: Forcings, Feedbacks, and the Climate Benefits of Forests', *Science*, CCCXX/320 (2008), pp. 1444–9

Bond, William J., 'Ancient Grasslands at Risk', *Science*, CCCLI/6269 (2016), pp. 120–22

—, and Guy Midgley, 'Carbon Dioxide and the Uneasy Interactions of Trees and Savannah Grasses', *Philosophical Transactions of the Royal Society of London B: Biological Sciences*, CCCLXVII/1588 (2012), pp. 601–12

—, and Catherine L. Parr, 'Beyond the Forest Edge: Ecology, Diversity and Conservation of the Grassy Biomes', *Biological Conservation*, CXLIII/10 (2010), pp. 2395–404

Bosch, J. M., and J. D. Hewlett, 'A Review of Catchment Experiments to Determine the Effect of Vegetation Changes on Water Yield and Evapotranspiration', *Journal of Hydrology*, LV/1–4 (1982), pp. 3–23

Bowler, Peter J., *Evolution: The History of an Idea* (Berkeley and Los Angeles, CA, 2003)

—, *The Norton History of the Environmental Sciences* (New York, 1993)

Braham, Roscoe R., 'Formation of Rain: A Historical Perspective', in *Historical Essays on Meteorology, 1919–1995*, ed. James Fleming (Boston, MA, 1996)

Brescius, Moritz von, *German Science in the Age of Empire* (Cambridge, 2018)

Brooks, C.E.P., 'The Influence of Forests on Rainfall and Run-Off', *Royal Meteorological Society Quarterly Journal*, LIV/225 (1928), pp. 1–17

Brutsaert, Wilfried, *Hydrology: An Introduction* (Cambridge, 2005)

Buckley, Eve E., *Technocrats and the Politics of Drought and Development in Twentieth-Century Brazil* (Chapel Hill, NC, 2017)

Burton, Jim, 'Robert FitzRoy and the Early History of the Meteorological Office', *British Journal for the History of Science*, XIX/2 (1986), pp. 147–76

Calder, Ian R., 'Forests and Hydrological Services: Reconciling Public and Science Perceptions', *Land Use and Water Resources Research*, II (2002), pp. 1–12
Carey, Mark, *In the Shadow of Melting Glaciers: Climate Change and Andean Society* (New York, 2010)
Carrus, Giuseppe, et al., 'Go Greener, Feel Better? The Positive Effects of Biodiversity on the Well-Being of Individuals Visiting Urban and Peri-Urban Green Areas', *Landscape and Urban Planning*, CXXXIV (2015), pp. 221–8
Chazdon, Robin L., 'Beyond Deforestation: Restoring Forests and Ecosystem Services on Degraded Lands', *Science*, CCCXX (2008), pp. 1458–60
Chen, Shuoyu, et al., 'Quantifying the Impact of Large-Scale Afforestation on the Atmospheric Water Cycle during Rainy Season over the Chinese Loess Plateau', *Journal of Hydrology*, DCXIX (2023), article 129326
Coen, Deborah R., *Climate in Motion: Science, Empire and the Problem of Scale* (Chicago, IL, 2018)
Cook, Harold, *Matters of Exchange: Commerce, Medicine, and Science in the Dutch Golden Age* (New Haven, CT, 2007)
Cronon, William, *Changes in the Land: Indians, Colonists, and the Ecology of New England* (New York, 1983)
Crossley, Robert, 'Percival Lowell and the History of Mars', *Massachusetts Review*, XLI (2000), pp. 297–318
Cruz, Felipe, 'Napalm Colonization: Native Peoples in Brazil's Aeronautical Frontiers', *Hispanic American Historical Review*, CI/3 (2021), pp. 461–89
Curtin, Philip D., 'The End of the "White Man's Grave"? Nineteenth-Century Mortality in West Africa', *Journal of Interdisciplinary History*, XXI/1 (1990), pp. 63–88
Cushman, Gregory, *Guano and the Opening of the Pacific World: A Global Ecological History* (Cambridge, 2013)
—, 'Humboldtian Science, Creole Meteorology, and the Discovery of Human-Caused Climate Change in Northern South America', *Osiris*, XXVI (2011), pp. 19–44
Davin, Edouard L., and Nathalie de Noblet-Ducoudré, 'Climate Impact of Global-Scale Deforestation: Radiative versus Nonradiative Process', *Journal of Climate*, XXIII/1 (2010), pp. 97–112
Davis, Diana K., *The Arid Lands: History, Power, Knowledge* (Cambridge, MA, and London, 2016)
—, 'Restoring Roman Nature: French Identity and North African Environmental History', in *Environmental Imaginaries of the Middle East and North Africa*, ed. Diana K. Davis and Edmund Burke III (Athens, OH, 2011)
—, *Resurrecting the Granary of Rome: Environmental History and French Colonial Expansion in North Africa* (Athens, OH, 2007)
Dettelbach, Michael, 'Humboldtian Science', in *Cultures of Natural History*, ed. Nicholas Jardine, James A. Secord and Emma C. Spray (Cambridge, 1996)
Dodds, Gordon B., *Hiram Martin Chittenden: His Public Career* (Lexington, KY, 1973)

—, 'The Stream-Flow Controversy: A Conservation Turning Point', *American History*, LVI/1 (1961), pp. 59–69

Dörries, Matthias, 'The Politics of Atmospheric Sciences: "Nuclear Winter" and Global Climate Change', *Osiris*, XXVI (2011)

Drayton, Richard, *Nature's Government: Science, Imperial Britain, and the 'Improvement' of the World* (New Haven, CT, 2000)

Dugatkin, Lee Alan, *Mr Jefferson and the Giant Moose: Natural History in Early America* (Chicago, IL, 2009)

Edwards, Paul N., *A Vast Machine: Computer Models, Climate Data and the Warming of Global Politics* (Cambridge, MA, 2013)

Ellingson, Terry Jay, *The Myth of the Noble Savage* (Berkeley and Los Angeles, CA, 2001)

Ellison, David, and Chinwe Ifejika Speranza, 'From Blue to Green Water and Back Again: Promoting Tree Shrub and Forest-Based Landscape Resilience in the Sahel', *Science of the Total Environment*, DCCXXXIX (2020), pp. 1–14

Ellison, David, Martyn N. Futter and Kevin Bishop, 'On the Forest Cover-Water Yield Debate: From Demand to Supply Side Thinking', *Global Change of Biology*, XIII/3 (2011), pp. 806–20

Ellison, David, et al., 'Europe's Forest Sink Obsession', EGU General Assembly 2022, Vienna, Austria, 23–27 May 2022, EGU22-8950 (2020), https://doi.org/10.5194/egusphere-egu22-8950.

—, 'Governance Options for Addressing Changing Forest-Water Relations', in *Forest and Water on a Changing Planet: Vulnerability, Adaptation and Governance Opportunities: A Global Assessment Report*, ed. Irena F. Creed and Meine van Noordwijk (Vienna, 2018)

—, 'Trees Forests and Water: Cool Insights for a Hot World', *Global Environmental Change*, XLIII (2017), pp. 51–61

Endfield, Georgina, and David Nash, 'Drought, Desiccation and Discourse: Missionary Correspondence and Nineteenth-Century Climate Change in Central Southern Africa', *Geographical Journal*, CLXVIII/1 (2002), pp. 33–47

Fairhead, James, and Mellissa Leech, *Misreading the African Landscape: Society and Ecology in a Forest-Savanna Mosaic* (Cambridge, 1996)

Fleming, James, *Fixing the Sky: The Checkered History of Weather and Climate Control* (New York, 2012)

—, *Historical Perspectives on Climate Change* (New York and Oxford, 1998)

Friedman, Robert Marc, *Appropriating the Weather: Vilhelm Bjerknes and the Construction of a Modern Meteorology* (Ithaca, NY, and London, 1989)

—, 'Sverre Petterssen, the Bergen School and Forecasts of D-Day', *Proceedings of the International Commission on History of Meteorology*, I/1 (2004), pp. 1–9

Ganguly, Akash, et al., 'Extreme Local Recycling of Moisture via Wetlands and Forests in North-East Indian Subcontinent: A Mini-Amazon', *Scientific Reports*, XIII (2023), pp. 1–10

Garfield, Seth, 'The Brazilian Amazon and the Transnational Environment, 1940–1990', in *Nation-States and the Global Environment: New*

Approaches to International Environmental History, ed. Erika Marie Bsumek, David Kinkela and Mark Atwood Lawrence (New York, 2013)
Gascoigne, John, *Joseph Banks and the English Enlightenment: Useful Knowledge and Polite Culture* (Cambridge, 1994)
—, *Science in the Service of Empire: Joseph Banks, the British State, and the Uses of Science in the Age of Revolution* (Cambridge, 1998)
Gates, Warren E., 'The Spread of Ibn Khaldûn's Ideas on Climate and Culture', *Journal of the History of Ideas*, XXVIII/3 (1967), pp. 415–22
Gebrehiwot, Solomon Gebreyohannis, et al., 'The Nile Basin Waters and the West African Rainforest: Rethinking the Boundaries', *Wires Water*, VI/1 (2018), pp. 1–8
Gerbi, Antonello, *The Dispute of the New World: The History of a Polemic, 1750–1900* (Pittsburgh, PA, 1973)
Gilbert, Natasha, 'Green Space: A Natural High', *Nature*, DXXXI (2016), pp. 56–7
Glacken, Clarence, *Traces on the Rhodian Shore: Nature and Culture in Western Thought from Ancient Times to the End of the Eighteenth Century* (Berkeley and Los Angeles, CA, and London, 1976)
Golinski, Jan, *British Weather and the Climate of Enlightenment* (Chicago, IL, 2007)
—, 'Debating the Atmospheric Constitution: Yellow Fever and the American Climate', *Eighteenth-Century Studies*, XLIX/2 (2016), pp. 149–65
Gresh, James, and Jason Counter, 'In Pursuit of Ecological Forestry: Historical Barriers and Ecosystem Implications', *Frontiers in Forests and Global Change*, IV (2021), pp. 1–9
Grove, Richard, *Green Imperialism: Colonial Expansion, Tropical Island Edens and the Origins of Environmentalism, 1600–1860* (Cambridge, 1996)
Guha, Ramchandra, *The Unquiet Woods: Ecological Change and Peasant Resistance in the Himalaya* (Berkeley, CA, 1989)
Hamblin, Jacob, *Arming Mother Nature: The Birth of Catastrophic Environmentalism* (Oxford, 2013)
Harper, Kristine, *Make It Rain: State Control of the Atmosphere in Twentieth-Century America* (Chicago, IL, 2017)
—, *Weather by the Numbers: The Genesis of Modern Meteorology* (Cambridge, MA, and London, 2008)
—, and Ronald E. Doel, 'Environmental Diplomacy in the Cold War', in *Environmental Histories of the Cold War*, ed. J. R. McNeill and Corinna Unger (Cambridge, 2010)
Hayes, Kevin, *The Road to Monticello: The Life and Mind of Thomas Jefferson* (New York, 2008)
Hogeboom, Rick, 'The Water Footprint Concept and Water's Grand Environmental Challenges', *One Earth*, II/3 (2020), pp. 218–22
Hölzl, Richard, 'Historicizing Sustainability: German Scientific Forestry in the Eighteenth and Nineteenth Centuries', *Science as Culture*, XIX (2010)

Hulme, Mike, *Weathered: Cultures of Climate Change* (London, 2017)

—, *Why We Disagree about Climate Change: Understanding Controversy, Inaction and Opportunity* (Cambridge, 2009)

Ioannidis, John, 'Why Most Published Research Findings Are False', *Plos Medicine*, II/8 (2005), pp. 696–701

Jackson, Robert B., et al., 'Trading Water for Carbon with Biological Carbon Sequestration', *Science*, CCCX/5756 (2005), pp. 1944–7

Kupperman, Karen Ordahl, 'The Puzzle of the American Climate in the Early Colonial Period', *American Historical Review*, LXXXVII/5 (1982), pp. 1262–89

Lawrence, Deborah, et al., 'The Unseen Effects of Deforestation: Biophysical Effects on Climate', *Frontiers in Forests and Global Change*, V (2022), pp. 1–13

Legg, Stephen, 'Debating the Climatological Role of Forests in Australia, 1827–1949', in *Climate, Science, and Colonization: Histories from Australia and New Zealand*, ed. J. Beattie, M. Henry and E. O'Gorman (New York, 2014)

Lehmann, Phillip, 'Average Rainfall and the Play of Colors: Colonial Experience and Global Climate Data', *Studies in History and Philosophy of Science Part A*, LXX (2018), pp. 38–49

—, 'Whither Climatology? Brückner's Climate Oscillations, Data Debates, and Dynamic Climatology', *History of Meteorology*, VII (2015), pp. 49–69

Lewis, Jamie, 'Theodore Roosevelt's Cautionary Tale', *Forest History Today* (2005), pp. 53–7

Linton, Jamie, 'Is the Hydrologic Cycle Sustainable? A Historical–Geographical Critique of a Modern Concept', *Annals of the Association of American Geographies*, XCVIII/3 (2008), pp. 630–49

—, *What Is Water? The History of a Modern Abstraction* (Vancouver, 2010)

Locher, Fabien, and Jean-Baptiste Fressoz, *Les Révoltes du ciel: Une histoire du changement climatique, XVe–XXe siècle* (Paris, 2020)

—, 'Modernity's Frail Climate: A Climate History of Environmental Reflexivity', *Critical Inquiry*, XXXVIII/3 (2020), pp. 579–98

Lovejoy, Arthur O., *The Great Chain of Being* (Cambridge, MA, 1964)

McIntyre, Iain, 'From the Local to the Global and Back Again: The Rainforest Information Centre and Transnational Environmental Activism in the 1980s', in *The Transnational Activist: Transformations and Comparisons from the Anglo-World since the Nineteenth Century*, ed. Stefan Berger and Sean Scalmer (Cham, 2017)

McKittrick, Meredith, 'An Empire of Rivers: Climate Anxiety, Imperial Ambition, and the Hydropolitical Imagination in Southern Africa, 1919–1945', *Journal of Southern African Studies*, XLI/3 (2015), pp. 485–504

—, 'Making Rain, Making Maps: Competing Geographies of Water and Power in 19th-Century Southwestern Africa', *Journal of African History*, LVIII/2 (2017), pp. 187–212

—, 'Talking about the Weather: The Language of Environmental Crisis in South Africa, 1915–1945', *Environmental History* (2018), pp. 3–27

—, 'Theories of "Reprecipitation" and Climate Change in the Settler Colonial World', *History of Meteorology*, VIII (2017), pp. 74–94
McNeil, J. R., and Peter Engelke, *The Great Acceleration: An Environmental History of the Anthropocene since 1945* (Cambridge, MA, 2016)
Mahony, Martin, and Samuel Randalls, eds, *Weather, Climate and the Geographical Imagination: Placing Atmospheric Knowledge* (Pittsburgh, PA, 2020)
Makarieva, Anastassia M., and Victor G. Gorshkov, 'Biotic Pump of Atmospheric Moisture as Driver of the Hydrological Cycle on Land', *Hydrology and Earth System Sciences*, XI/2 (2007), pp. 1013–103
—, and Bai-Lan Li, 'Conservation of Water Cycle on Land via Restoration of Natural Closed-Canopy Forests: Implications for Regional Landscape Planning', *Ecological Research*, XXI/6 (2006), pp. 897–906
Marsh, George Perkins, *Man and Nature; or, Physical Geography as Modified by Human Actions* (New York, 1864)
Moore, Peter, *The Weather Experiment: The Pioneers Who Sought to See the Future* (London, 2015)
Morgan, Ruth, *Running Out? Water in Western Australia* (Perth, 2015)
Morris, Ellison, et al., 'Trees, Forests and Water', *Global Environmental Change*, XLIII (2017), pp. 51–61
Nebeker, Frederik, *Calculating the Weather: Meteorology in the 20th Century*, 1st edn (San Diego, CA, 1995)
Nicholson, Sharon E., *Dryland Climatology* (Cambridge, 2011)
Paul, Supantha, et al., 'Weakening of Indian Summer Monsoon Rainfall Due to Changes in Land Use Land Cover', *Scientific Reports*, VI (2016), pp. 1–10
Pedgley, David E., 'The British Rainfall Organization, 1859–1919', *Special Issue: The 150th Anniversary of British Rainfall Organization*, LXV/5 (2010), pp. 115–17
Pittock, Jamie, 'Australian Climate, Energy and Water Policies: Conflicts and Synergies', *Australian Geographer*, XLIV/1 (2013), pp. 3–22
Purcell, Terry, Erminielda Peron and Rita Berto, 'Why Do Preferences Differ between Scene Types?', *Environment and Behaviour*, XXXIII/1 (2001), pp. 93–106
Rand, Lisa, 'Falling Cosmos: Nuclear Re-Entry and the Environmental History of Earth Orbit', *Environmental History*, IV/1 (2019), pp. 78–103
Reich, Peter, 'Biogeochemistry: Taking Stock of Forest Carbon', *Nature Climate Changes*, I (2011), pp. 346–7
Riley, James C., *Rising Life Expectancy: A Global History* (Cambridge, 2001)
Rogers, Thomas, *The Deepest Wounds: A Labor and Environmental History of Sugar in Northeast Brazil* (Chapel Hill, NC, 2010)
Rudel, Thomas, *Tropical Forests: Regional Paths of Destruction and Regeneration in the Late Twentieth Century* (New York, 2005)
Rupke, Nicolaas A., *Alexander von Humboldt: A Metabiography* (Chicago, IL, 2008)
Saberwal, V. K., 'Science and the Desiccationist Discourse of the 20th Century', *Environment and History*, IV/3 (1998), pp. 309–43
Saito, Kohei, *Marx in the Anthropocene: Towards the Idea of Degrowth Communism* (Cambridge, 2023)

Saladanha, Indra Munshi, 'Colonialism and Professionalism: A German Forester in India', *Environment and History*, II/2 (1996), pp. 195–219

Salsburg, David, *The Lady Tasting Tea: How Statistics Revolutionized Science in the Twentieth Century* (New York, 2001)

Sanderson, Michael, et al., 'Influences of EU Forests on Weather Patterns: Final Report', *Meteorological Office Hadley Centre* (2012), pp. 1–32

Secord, James, *Victorian Sensation: The Extraordinary Publication, Reception, and Secret Authorship of 'Vestiges of the Natural History of Creation'* (Chicago, IL, 2003)

Sheil, Douglas, 'How Plants Water Our Planet: Advances and Imperatives', *Trends in Plant Science*, XIX/4 (2014), pp. 209–11

——, and Daniel Murdiyarso, 'How Forests Attract Rain: An Examination of a New Hypothesis', *Bioscience*, LIX/4 (2009), pp. 341–7

Showers, Kate, *Imperial Gullies: Soil Erosion and Conservation in Lesotho* (Athens, OH, 2005)

Sivaramakrishnan, Kalyanakrishnan, *Modern Forests: Statemaking and Environmental Change in Colonial Eastern India* (Stanford, CA, 1999)

Smith, Chris, Jessica C. A. Baker and Dominick Spracklen, 'Tropical Deforestation Causes Large Reductions in Observed Precipitation', *Nature*, DCXV (2023), pp. 270–75

Stehr, Nico, and Hans von Storch, eds, *Eduard Brückner: The Sources and Consequences of Climate Change and Climate Variability in Historical Times* (Dordrecht, 2000)

Teo, Hoong Chen, et al., 'Large-Scale Reforestation Can Increase Water Yield and Reduce Drought Risk for Water-Insecure Regions in the Asia-Pacific', *Global Change Biology*, XXVIII/21 (2022), pp. 6385–40

Thompson, Kenneth, 'Forests and Climate Change in America: Some Early Views', *Climatic Change*, III/1 (1980), pp. 47–64

Tyrrell, Ian, *Crisis of the Wasteful Nation: Empire and Conservation in Theodore Roosevelt's America* (Chicago, IL, 2015)

Van den Berg, Agnes, Terry Hartig and Henk Staats, 'Preference for Nature in Urbanized Societies: Stress, Restoration and the Pursuit of Sustainability', *Journal of Social Sciences,* LXIII/1 (2007), pp. 79–96

Van der Ent, Ruud J., et al., 'Origin and Fate of Atmospheric Moisture over Continents', *Water Resources Research*, XLVI/9 (2010), pp. 1–12

——, 'The Importance of Proper Hydrology in the Forest Cover–Water Yield Debate: Commentary on Ellison et.al.', *Global Change Biology*, XVIII/9 (2012), pp. 2677–80

Van Dijke, et al., 'Shifts in Regional Water Availability Due to Global Tree Restoration', *Nature Geosciences*, XV (2022), pp. 363–8

Van Sittert, Lance, '"Our Irrepressible Fellow-Colonist": The Biological Invasion of Prickly Pear (*Opuntia ficus-indica*) in the Eastern Cape, *c.* 1890–*c.* 1910', *Journal of Historical Geography*, XXVIII/3 (2002), pp. 397–410

Walker, Malcolm, *History of the Meteorological Office* (Cambridge, 2011)

Walls, Laura Dassow, *The Passage to Cosmos: Alexander von Humboldt and the Shaping of America* (Chicago, IL, 2009)
Wang, Feng, et al., 'Vegetation Restoration in Northern China: A Contrasted Picture', *Land Degradation and Development*, XXXI/6 (2019), pp. 669–76
Warde, Paul, Libby Robin and Sverker Sörlin, *The Environment: A History of the Idea* (Baltimore, MD, 2018)
Weart, Spencer, *The Discovery of Global Warming* (Cambridge, MA, 2008)
Weintraub, David, *Life on Mars: What to Know before We Go* (Princeton, NJ, 2018)
White, Lynn Jr, 'The Historical Roots of Our Ecological Crisis', *Science New Series*, CLV/3767 (1967), pp. 1203–7.
Williams, Michael, *Americans and Their Forests: A Historical Geography* (Cambridge, 1992)
—, *Deforesting the Earth: From Prehistory to Global Crisis* (Chicago, IL, 2006)
Wood, Charles, 'Environmental Hazards, Eighteenth-Century Style', in *Old World, New World: America and Europe in the Age of Jefferson*, ed. Leonard J. Sadosky, Peter Nicolaisen, Peter S. Onuf and Andrew J. O'Shaughnessy (Monticello, 2010)
Wood, Gordon S., 'America's First Climate Debate: Thomas Jefferson Questioned the Science of European Doomsayers', *American History*, XL/6 (2010), pp. 58–63
Worster, Donald, *Dust Bowl: The Southern Plains in the 1930s* (New York, 1979/2004)
Wulf, Andrea, *The Invention of Nature: The Adventures of Alexander von Humboldt: The Lost Hero of Science* (London, 2015)
Zilberstein, Anya, *A Temperate Empire: Making Climate Change in Early America* (Oxford, 2016)

ACKNOWLEDGEMENTS

This book is dedicated to the late Frederick J. Kruger, a dear friend and a great ecological thinker. Fred inspired deep conversations not only on science but on philosophy and how to live life.

We drafted the book while working at Western Sydney University, the University of Johannesburg and the Australian National University. Special thanks go to our colleagues in history at Western Sydney and Johannesburg, where we spent the bulk of our time when writing the book. Our highly supportive Deans, Matt McGuire, Kamilla Naidoo, Alex Broadbent and Peter Hutchings, allowed us to foster our interdisciplinary passions. The idea for the book started at the Australian National University and we received funding from the Australian Research Council (DP110104024). We are especially grateful for the support provided by the Centre for Environmental History and history colleagues at ANU. Moreover, the book reflects our engagements with many scholars. Scholars who generously helped us at various times include, but are not limited to (in no particular order), James Beattie, Ruth Morgan, Simon Pooley, Alexandro Antonello, Andrea Gaynor, Fredrik Albritton Jonsson, Diana Davis, Libby Robin, Tom Griffiths, Lance van Sittert, Sandra Swart, Jane Carruthers, Fei Sheng, Joseph Hodge, Ulrike Kirchberger, Adrian Howkins, Shoko Mizuno, Paul Munro, Taro Takemoto, Togo Tsukahara, Koji Nakashima, Adam Lucas, Ian Tyrrell, Vinita Damodaren, Paul Munro, Steven Anderson and Jamie Lewis.

We are grateful to scientific input from a number of scholars. David Ellison generously provided insights on supply-side hydrology by reading chapters of the book, especially the penultimate chapter, which explores recent developments in research studying the link between forests and atmospheric moisture recycling. In South Africa, David Richardson, Laurence Kruger, William Bond, Guy Midgley and Ross Shackleton, among many others, provided ideas, housing, a sabbatical and many dinners that expanded our knowledge of hydrology, forestry, ecology and climate change. In Australia, the Hawkesbury Institute for the Environment (especially Mark Tjoelker) and the Institute for Culture and Society (especially Paul James) supported our interdisciplinary research on forests and climate. In China, Klaus von Gadow and Beijing Forestry University invited us to visit China along with a group of foresters. That trip led to the publication of 'The Enduring Link between Forest Cover and Rainfall: A Historical Perspective on Science and Policy Discussions', *Forest Ecosystems*, v/5 (2018).

Thanks to Reaktion Books, especially Michael Leaman and Alex Ciobanu for supporting us while we drafted the book. Michael provided valuable commentary that helped us to revise the manuscript. Two helpful and learned anonymous reviewers helped us to revise the book to better communicate with popular and scholarly audiences.

Finally, we want to acknowledge our families for supporting us. A special thanks goes to Michael Bennett, who kindly helped us to edit the manuscript for readability based on his decades of working in journalism. Ina Barton, an author in her own right, inspired us to complete the book so she could see it for herself.

INDEX